THE POCKET GUIDE TO
ASTRONOMY

THE POCKET GUIDE TO
ASTRONOMY

IAN RIDPATH
with star maps by
WIL TIRION

DRAGON'S
WORLD

Dragon's World Ltd
Limpsfield
Surrey RH8 0DY
Great Britain

First published by Dragon's World 1990

British Library Cataloguing in Publication Data

Ridpath, Ian
  The pocket guide to astronomy.
  1. Astronomy
  I. Title
  520

ISBN 1-85028-106-8

Editors: John Woodruff, Michael Downey
Design: David Hunter
Editorial Director: Pippa Rubinstein

Typeset by The Works, Devon, England
Printed in Singapore

Contents

INTRODUCING THE NIGHT SKY

Our view of the heavens is inevitably affected by the planet on which we live. Every day the Earth spins once on its axis, so that celestial objects appear to move across the sky from east to west. This accounts for the rising and setting of the Sun that gives us day and night, as well as the rising and setting of the Moon and stars during each night.

Your view of the sky is also affected by your position on Earth. To appreciate this, imagine that you are standing at the north pole on Earth. From there, the north pole star, Polaris, lies directly overhead. Only the stars in the northern celestial hemisphere are visible to you, and all of them appear to circle around Polaris once a day as the Earth turns on its axis. If you then moved from the pole towards the equator you would see Polaris gradually become lower in the sky until, when you reached the equator, it would lie on the northern horizon.

As seen from the Earth's north pole, stars do not rise or set but circle parallel to the horizon. Polaris lies overhead.

It happens that the altitude of Polaris above the northern horizon is equal to your latitude on Earth (or almost so, since Polaris lies about one degree away from the exact pole). For example, from latitude 40°N Polaris is approximately 40° above the northern horizon, from latitude 50°N it is approximately 50° above the northern horizon, and so on. As shown opposite, stars close to Polaris never dip below the horizon; they are always visible, circling around the pole, and hence are known as *circumpolar* stars. Just how much of the sky is circumpolar depends on your latitude. From the north pole, for example, all stars visible are circumpolar, but the area of sky that is circumpolar becomes progressively smaller as you approach the equator. From the equator itself, no stars are circumpolar – they all rise and set.

In addition to its daily spin on its axis the Earth follows its year-long orbit around the Sun, but this has a much less perceptible effect on the appearance of the sky. Watch the eastern horizon over several nights and you will find that the rising time of stars gets progressively earlier – by about four minutes a night, in fact (and, equally, the setting time of stars at the western horizon gets progressively earlier). Hence the stars visible in winter are different from those we see in the summer. After a year the Earth will have completed one orbit around the Sun and the stars will be back where they were one year before.

As a result of the Earth's movement around its orbit, the

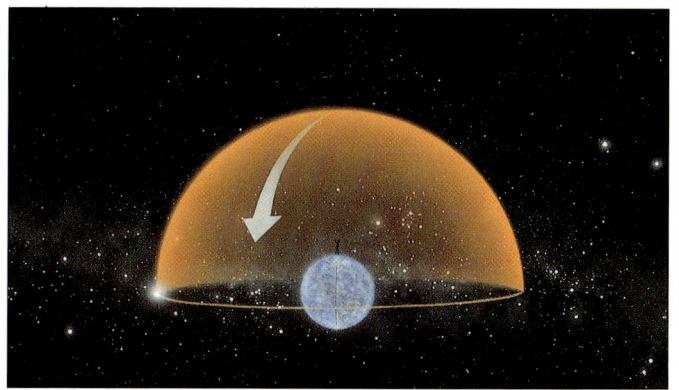

From the Earth's equator, an observer sees all stars rise and set as the Earth turns. Polaris lies on the northern horizon.

Sun appears to follow a path known as the *ecliptic* (because eclipses happen when the Moon lies near it). Some stars are invisible at certain times of the year because they are in the daytime sky. The constellations that lie along the ecliptic are known collectively as the *zodiac*.

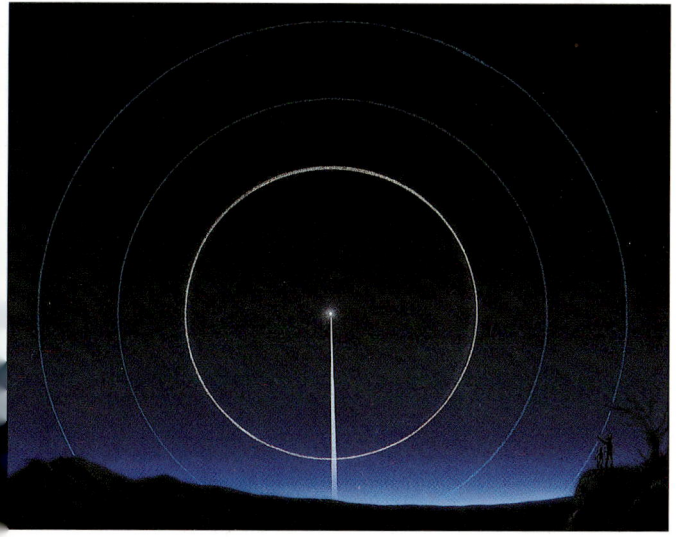

The angle of Polaris above your northern horizon (indicated by the vertical line) is equal to your latitude. Stars closer than this to Polaris never set – they are circumpolar (inner circle).

POSITIONS IN THE SKY. It is convenient to imagine that all stars lie on the surface of an imaginary sphere surrounding the Earth. This sphere is called the *celestial sphere*. The poles of the celestial sphere lie directly above the poles of the Earth, and the celestial equator lies directly above the Earth's equator. The ecliptic lies at an angle of about 23.5° to the celestial equator; this is because the Earth's axis is tilted at 23.5° to the vertical.

In the same way that positions on the Earth's surface are identified by their longitude and latitude, stars and other celestial objects are located by their coordinates on the celestial sphere. The celestial equivalent of longitude is *right ascension*, or RA. It is measured from 0 to 24 hours, since the celestial sphere appears to rotate once every 24 hours as the Earth spins. The zero point of right ascension lies where the Sun crosses the celestial equator into the northern hemisphere each spring; it is commonly called the *vernal* (or *spring*) *equinox*. The celestial equivalent of latitude is *declination*, or dec., and like latitude on Earth it is measured from 0° at the equator to 90° at the poles.

Distances on the celestial sphere are measured as angles. As an example, angular distances between the stars of the

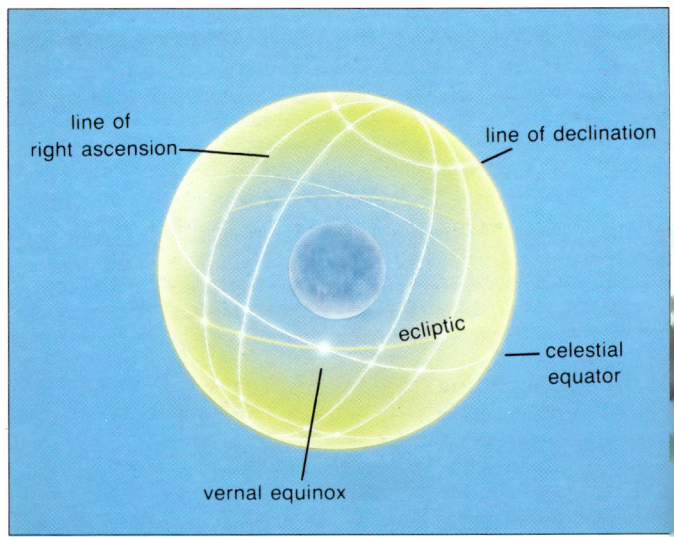

The celestial sphere – an imaginary sphere surrounding the Earth. The Sun's path, the ecliptic, lies at 23.5° to the celestial equator and cuts it at the vernal equinox.

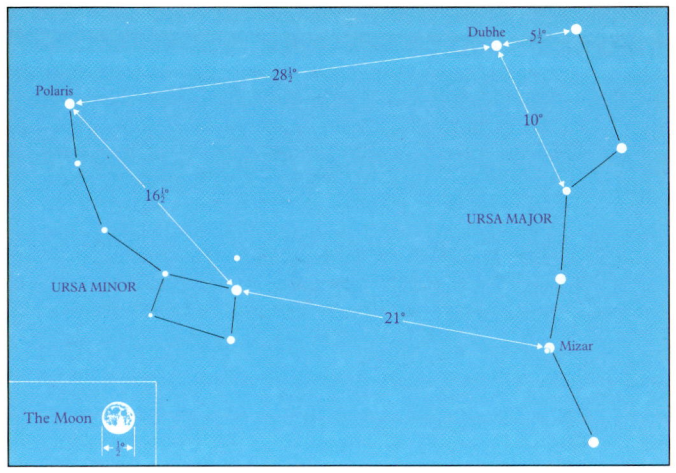

Angular distances in the sky.

Plough and Ursa Minor are shown in the diagram. Angles smaller than a degree are measured in arc minutes, one arc minute being one-sixtieth of a degree. For example, the diameters of the Moon and Sun are each about half a degree, or 30 arc minutes. The unaided human eye can distinguish objects as close as 2 or 3 arc minutes apart.

The smallest angles of all, such as the separation between two close stars or the apparent diameter of a planet, are measured in arc seconds, often abbreviated *arcsec*. An arc second is one-sixtieth of an arc minute, and corresponds roughly to the width of a small coin observed at at distance of 4 km. Jupiter has the largest apparent diameter of any planet, about 50 arcsec when at its closest to us.

Something happens on the celestial sphere that does not happen on Earth: the coordinates of celestial objects are slowly but steadily changing, by about a degree every 70 years. This effect is called *precession*, and is caused by the Earth slowly wobbling in space like a spinning top. Over a period of nearly 26,000 years each of the Earth's poles traces out a complete circle on the celestial sphere, and during the same time the zero point of right ascension makes one circuit of the celestial sphere. Catalogues therefore list star positions for a fixed date, or *epoch*. The epoch currently used is the start of the year 2000, and the maps in this book are drawn with star positions for that date – although in practice the differences are negligible for many decades either side of the year 2000.

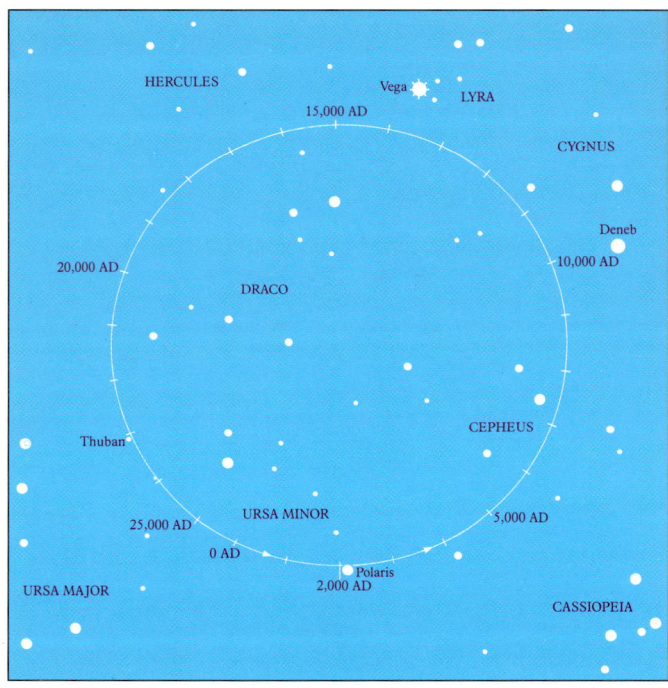

Precession causes the Earth's north pole to trace out a circle on the celestial sphere once every 26,000 years.

Incidentally, in the days of the Ancient Greek astronomers the zero point of right ascension lay in the constellation Aries. Precession has now moved it into neighbouring Pisces, but the zero point is still sometimes referred to as the *first point of Aries*.

One consequence of precession is that the pole star is constantly changing. Polaris will be at its closest to the north celestial pole around the year 2100, and will then start to move away again. We are fortunate to live at a time when a moderately bright star lies near the north celestial pole (see the diagram above). (It is a popular misconception that the north pole star is exceptionally bright.)

SEASONS are a consequence of the axial tilt of the Earth. Without that tilt, the Sun's path would coincide with the celestial equator and there would be no seasons. As mentioned above, the Sun's path – the ecliptic – is a circle inclined at 23.5° to the celestial equator.

Twice a year the Sun lies exactly on the celestial equator at the points where the ecliptic cuts it, on March 21 and

September 23. These dates are known as the *equinoxes*, when day and night are virtually equal in length the world over.

After the *vernal equinox* (or spring equinox) in March, the Sun moves north of the celestial equator. It reaches its most northerly point, the *summer solstice*, around June 21, when the northern hemisphere experiences its longest day. On that date the Sun is directly overhead at noon on the Tropic of Cancer, latitude 23.5° north.

The Sun then moves back towards the celestial equator, crossing it southbound at the *autumnal equinox* in September, and then moving into the southern half of the sky. The Sun reaches its most southerly point at the *winter solstice* around December 22, the shortest day in the northern hemisphere. On this date the Sun is directly overhead at noon as seen from the Tropic of Capricorn, latitude 23.5° south.

STAR BRIGHTNESSES are measured in *magnitudes*, a system invented by the Ancient Greeks 2000 years ago and refined by modern astronomers. The Greeks divided the stars into six classes, sixth-magnitude stars being the faintest visible to the naked eye and first-magnitude stars being the brightest. On this scale Polaris is a second-magnitude star.

When in the last century astronomers developed instruments to measure star brightnesses accurately, the magnitude scale was redefined so that a difference of five magnitudes exactly equals a brightness difference of 100 times. Hence a star of magnitude 1.0 is 100 times brighter than one of magnitude 6.0. Once the magnitude scale had been defined in this way it was necessary to modify it to include the brightest stars, which are more than 100 times brighter than the faintest ones visible. So the magnitude scale now continues up through magnitude 0 and into negative magnitudes for the very brightest objects. Vega, a brilliant star of summer, has a magnitude of almost exactly 0, while Sirius, the brightest star in the sky, is magnitude −1.46.

The magnitude scale is used for all celestial objects, not just stars. The planets Venus and Jupiter (magnitudes respectively −4.7 and −2.9 at best) are even brighter than Sirius, and sometimes so is Mars (−2.8 at best). In fact, the magnitude scale can be extended indefinitely in both directions, up to the very brightest objects such as the Moon and Sun (magnitudes −12.7, when full, and −26.8 respectively) and down to the faintest objects visible through large telescopes, which are around magnitude 24 (the magnitude is assumed to be positive if no sign is given). With modern instruments, astronomers can measure brightnesses to a precision of a hundredth or even a thousandth of a magnitude.

The faintest magnitude visible is known as the *limiting magnitude*. While the nominal naked-eye limit is sixth magnitude, this is true only under clear, dark skies away from haze and streetlights. The actual limiting magnitude on a given occasion depends on the sky conditions and also on the observer's eyesight. In the polluted air of towns, the limiting magnitude may be no better than fourth magnitude. Under ideal conditions observers with exceptional eyesight have reported seeing stars of seventh magnitude. Much fainter stars can be seen through binoculars and telescopes (see page 30).

Objects near the horizon look much dimmer because their light has passed through a greater thickness of the Earth's atmosphere. This effect, called *extinction*, amounts to about one magnitude at 10° from the horizon, and even more for

TOP TWENTY BRIGHTEST STARS

Star	Name	Coordinates RA (2000.0) h m	Dec. ° ′	Mag.
α Canis Majoris	Sirius	06 45.1	− 16 43	− 1.46
α Carinae	Canopus	06 24.0	− 52 42	− 0.72
α Centauri	Rigil Kentaurus	14 39.6	− 60 50	− 0.27 [a]
α Boötis	Arcturus	14 15.7	+ 19 11	− 0.04
α Lyrae	Vega	18 36.9	+ 38 47	+ 0.03
α Aurigae	Capella	05 16.7	+ 46 00	0.08
β Orionis	Rigel	05 14.5	− 08 12	0.12
α Canis Minoris	Procyon	07 39.3	+ 05 13	0.38
α Eridani	Achernar	01 37.7	− 57 14	0.46
α Orionis	Betelgeuse	05 55.2	+ 07 24	0.50 [b]
β Centauri	Hadar	14 03.8	− 60 22	0.61
α Aquilae	Altair	19 50.8	+ 08 52	0.77
α Crucis	Acrux	12 26.6	− 63 06	0.79 [a]
α Tauri	Aldebaran	04 35.9	+ 16 31	0.85 [b]
α Scorpii	Antares	16 29.4	− 26 26	0.96 [b]
α Virginis	Spica	13 25.2	− 11 10	0.98
β Geminorum	Pollux	07 45.3	+ 28 01	1.14
α Piscis Austrini	Fomalhaut	22 57.6	− 29 37	1.16
α Cygni	Deneb	20 41.4	+ 45 17	1.25
β Crucis	Mimosa	12 47.7	− 59 41	1.25

[a] Combined magnitude of double star.
[b] Average magnitude of variable star.

Data from the Yale *Bright Star Catalogue*, 4th edition.

objects closer to the horizon. Therefore it is always best to observe objects when they are high in the sky.

CONSTELLATIONS. We have inherited from the Ancient Greeks not only the magnitude scale but also their division of the sky into star groups known as *constellations*. Originally, the Greek constellations were star patterns that represented characters from mythology. To modern astronomers, a constellation is simply an area of sky whose boundaries are laid down by the International Astronomical Union, astronomy's governing body, although many of the ancient names have been retained.

The 48 constellations known to the Greeks have been extended to 88 as astronomers have filled in gaps between the Greek figures and extended their mapping to the southern parts of the sky that were below Greek horizons. Nowadays, every part of the sky belongs to one of the 88 constellations (see pages 16 and 17).

Most of the stars in a constellation have no physical connection with one another, but simply happen to lie in roughly the same direction as seen from Earth. Because of the haphazard way in which the constellation system has arisen, constellations vary widely in size and shape. Hydra, the largest constellation, covers an area of sky 19 times greater than the smallest, Crux.

STAR NAMES. In 1603 Johann Bayer, a German astronomer, assigned Greek letters to the brightest stars in each constellation, usually (but not always) in order of decreasing brightness. His system has been followed to this day, so that stars are often referred to by names such as γ Andromedae, ζ Aquarii and β Canis Minoris. Note that the genitive form of the constellation's name is always used when referring to stars in this way. For example, γ Andromedae means 'gamma of Andromeda'.

Some stars also have proper names, such as Sirius, Betelgeuse and Canopus. These names come from a mixture of sources. Some are Greek in origin and others

THE GREEK ALPHABET	
α alpha	ν nu
β beta	ξ xi
γ gamma	o omicron
δ delta	π pi
ε epsilon	ϱ rho
ζ zeta	σ sigma
η eta	τ tau
θ theta	υ upsilon
ι iota	φ phi
κ kappa	χ chi
λ lambda	ψ psi
μ mu	ω omega

continued on page 18

THE CONSTELLATIONS

Name	Genitive	Abbr.	Area (square degrees)	Order of size
Andromeda	Andromedae	And	722	19
Antlia	Antliae	Ant	239	62
Apus	Apodis	Aps	206	67
Aquarius	Aquarii	Aqr	980	10
Aquila	Aquilae	Aql	652	22
Ara	Arae	Ara	237	63
Aries	Arietis	Ari	441	39
Auriga	Aurigae	Aur	657	21
Boötes	Boötis	Boo	907	13
Caelum	Caeli	Cae	125	81
Camelopardalis	Camelopardalis	Cam	757	18
Cancer	Cancri	Cnc	506	31
Canes Venatici	Canum Venaticorum	CVn	465	38
Canis Major	Canis Majoris	CMa	380	43
Canis Minor	Canis Minoris	CMi	183	71
Capricornus	Capricorni	Cap	414	40
Carina	Carinae	Car	494	34
Cassiopeia	Cassiopeiae	Cas	598	25
Centaurus	Centauri	Cen	1060	9
Cepheus	Cephei	Cep	588	27
Cetus	Ceti	Cet	1231	4
Chamaeleon	Chamaeleontis	Cha	132	79
Circinus	Circini	Cir	93	85
Columba	Columbae	Col	270	54
Coma Berenices	Comae Berenices	Com	386	42
Corona Australis	Coronae Australis	CrA	128	80
Corona Borealis	Coronae Borealis	CrB	179	73
Corvus	Corvi	Crv	184	70
Crater	Crateris	Crt	282	53
Crux	Crucis	Cru	68	88
Cygnus	Cygni	Cyg	804	16
Delphinus	Delphini	Del	189	69
Dorado	Doradus	Dor	179	72
Draco	Draconis	Dra	1083	8
Equuleus	Equulei	Equ	72	87
Eridanus	Eridani	Eri	1138	6
Fornax	Fornacis	For	398	41
Gemini	Geminorun	Gem	514	30
Grus	Gruis	Gru	366	45
Hercules	Herculis	Her	1225	5
Horologium	Horologii	Hor	249	58
Hydra	Hydrae	Hya	1303	1
Hydrus	Hydri	Hyi	243	61
Indus	Indi	Ind	294	49

Name	Genitive	Abbr.	Area (square degrees)	Order of size
Lacerta	Lacertae	Lac	201	68
Leo	Leonis	Leo	947	12
Leo Minor	Leonis Minoris	LMi	232	64
Lepus	Leporis	Lep	290	51
Libra	Librae	Lib	538	29
Lupus	Lupi	Lup	334	46
Lynx	Lyncis	Lyn	545	28
Lyra	Lyrae	Lyr	286	52
Mensa	Mensae	Men	153	75
Microscopium	Microscopii	Mic	210	66
Monoceros	Monocerotis	Mon	482	35
Musca	Muscae	Mus	138	77
Norma	Normae	Nor	165	74
Octans	Octantis	Oct	291	50
Ophiuchus	Ophiuchi	Oph	948	11
Orion	Orionis	Ori	594	26
Pavo	Pavonis	Pav	378	44
Pegasus	Pegasi	Peg	1121	7
Perseus	Persei	Per	615	24
Phoenix	Phoenicis	Phe	469	37
Pictor	Pictoris	Pic	247	59
Pisces	Piscium	Psc	889	14
Piscis Austrinus	Piscis Austrini	PsA	245	60
Puppis	Puppis	Pup	673	20
Pyxis	Pyxidis	Pyx	221	65
Reticulum	Reticuli	Ret	114	82
Sagitta	Sagittae	Sge	80	86
Sagittarius	Sagittarii	Sgr	867	15
Scorpius	Scorpii	Sco	497	33
Sculptor	Sculptoris	Scl	475	36
Scutum	Scuti	Sct	109	84
Serpens	Serpentis	Ser	637	23
Sextans	Sextantis	Sex	314	47
Taurus	Tauri	Tau	797	17
Telescopium	Telescopii	Tel	252	57
Triangulum	Trianguli	Tri	132	78
Triangulum Australe	Trianguli Australis	TrA	110	83
Tucana	Tucanae	Tuc	295	48
Ursa Major	Ursae Majoris	UMa	1280	3
Ursa Minor	Ursae Minoris	UMi	256	56
Vela	Velorum	Vel	500	32
Virgo	Virginis	Vir	1294	2
Volans	Volantis	Vol	141	76
Vulpecula	Vulpeculae	Vul	268	55

Latin, but most are Arabic, since Arab astronomers preserved the Greek tradition during the Dark Ages in Europe. Numerous star names were taken from Arab sources after the Renaissance, but they were often written wrongly so that many of the apparently Arabic names are actually meaningless.

Only the brighter stars have proper names and Greek letters. The rest are labelled by their number in one of many catalogues. Frequently used are *Flamsteed numbers*, which were given to the 3000 stars catalogued by the first Astronomer Royal of England, John Flamsteed (although the numbers were assigned not by Flamsteed himself, but by other astronomers after his death). Stars with Greek letters can also have Flamsteed numbers – for example, α Tauri (Aldebaran) has the Flamsteed number 87 Tauri – but usually the numbers are used only for stars with no Greek letter.

Even so, many stars do not have a Flamsteed number. Other catalogue numbers frequently encountered are those from the *Harvard Revised Photometry* (HR), the *Henry Draper Catalogue* (HD) and the *Smithsonian Astrophysical Observatory Star Catalog* (SAO). Many stars with special characteristics, notably variable stars, also have their own designations.

STAR DISTANCES are measured not in kilometres, but in light years or parsecs. A *light year* is the distance that a beam of light travels in one calendar year. Light travels at the fastest speed in the Universe, nearly 300,000 km per second, but still requires 4.3 years to reach us from the nearest star to the Sun, α Centauri. (Strictly speaking, a small, faint companion star to α Centauri called Proxima Centauri is about 0.1 light year closer, but it is not bright enough to be visible to the naked eye.) For comparison, the Moon is just over 1 light second away, the Sun is 8.3 light minutes away, and the planet Neptune is over 4 light hours distant.

Professional astronomers prefer another unit of distance, the *parsec*, which is approximately 3.26 light years. The parsec originates in the way in which distances to the nearest stars are measured, by a process of triangulation similar to that used by surveyors. Astronomers measure star positions at intervals during the year as the Earth orbits the Sun. Stars which are relatively close to us show microscopic changes in position when viewed against the background of more distant stars from different places around the Earth's orbit. This shift in position is too small to be seen with the naked eye, but it can be detected on photographs taken with large telescopes.

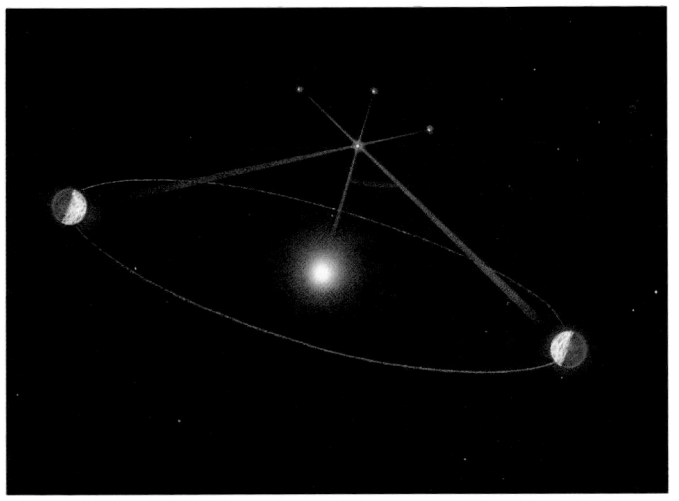

A star's parallax (the angle marked) is measured by observing it from opposite sides of the Earth's orbit.

The displacement of a star from its average position is known as its *parallax*, and is measured in seconds of arc (or arc seconds – see page 11). The smaller the parallax, the more distant the star. A parallax of one *sec*ond of arc would correspond to an object one parsec away, if the parallax were 0.1 arcsec the object would be ten parsecs away, and so on. The star nearest the Sun, α Centauri, has the largest stellar parallax, 0.75 arcsec.

The smallest parallax that can be measured reliably from Earth is about 0.04 arcsec, equivalent to a distance of 25 parsecs. For more distant stars we have to estimate their distances by comparing their calculated light output with their observed magnitude. This procedure is fraught with difficulties, and many published distances are unreliable. Deneb, in the constellation Cygnus, is the most distant of the bright stars visible in the night sky, an estimated 2000 light years away.

STARS AND PLANETS. Stars move across the sky during the night as the Earth spins, but they do not change position relative to each other – at least, not noticeably in a human lifetime. The constellation patterns we see today have changed little from those visible in Ancient Greek times. What changes there have been are caused by the slight motions of the stars themselves, imperceptible to the naked eye except over centuries. (A *shooting star*, by the way, has nothing to do with real stars. It is a piece of cosmic dust burning up in the Earth's atmosphere, and is properly known as a *meteor* – see page 96.)

However, there are bright, starlike objects that do move noticeably from night to night. These are the *planets*, bodies in orbit around the Sun. Including the Earth, there are nine planets orbiting the Sun, and they shine by reflecting the Sun's light. As they move across the celestial sphere they keep close to the ecliptic, so if you see a bright 'star' that disturbs the familiar pattern of the constellations of the zodiac, it will be one of the planets.

Five planets are visible to the naked eye: in order of distance from the Sun they are Mercury, Venus, Mars, Jupiter and Saturn. The brightest of them all is Venus. Mercury, being so close to the Sun, can be difficult to spot. Two more planets, Uranus and Neptune, are visible in binoculars. Pluto, at the edge of the Solar System, is so small and faint that it needs a sizable telescope to be seen.

MILKY WAY. Across the sky on clear, dark nights stretches a misty band of light known as the Milky Way. Through binoculars or small telescopes the Milky Way breaks up into a mass of faint stars. All these stars are distant members of our Galaxy, a spiral-shaped mass of stars of which the Sun is a member. The stars that make up the familiar constellation patterns are also members of our Galaxy, only much closer to us than those in the Milky Way. Our Galaxy is about 100,000 light years in diameter. Binoculars and telescopes show other far-off galaxies similar to our own (see page 127).

The Space Shuttle Challenger on its last successful flight in November 1985 leaves a bright trail (partly obscured by cloud).

ARTIFICIAL SATELLITES. From time to time on a clear night you may see what looks like a bright star gliding across the heavens, taking a few minutes to cross from one horizon to the other. This will be an artificial satellite, orbiting the Earth at a height of a few hundred kilometres. Most satellites that can be seen with the naked eye take around 90 or 100 minutes to complete one orbit of the Earth. Do not confuse them with high-flying aircraft; if you are uncertain, the navigation lights of an aircraft will show up in binoculars, whereas a satellite will still appear as a starlike point.

Satellites shine because they reflect the Sun's light, and the brightness of a satellite depends on both its size and its altitude. Large objects in low orbits, such as the Space Shuttle or the Soviet Space Station Mir, can appear brighter than the brightest stars. Many satellites move from west to east, but others have orbits passing over the polar regions, and can be seen moving from north to south or vice versa. The term 'satellite' is applied to any orbiting object, including discarded rocket stages and pieces of debris. There are thousands of man-made objects of all sizes orbiting the Earth. Dozens may be seen with the naked eye on a good night, while hundreds more are visible in binoculars.

Satellites can fade from view as they cross the sky, or appear as if from nowhere. This is because they are entering or leaving the Earth's shadow – that is, moving into or out of eclipse. Other satellites, particularly rocket stages, flash because they are tumbling in orbit. Occasionally you may see a point-like flash lasting a second or so with no apparent cause. Such flashes are becoming increasingly common and are due to sunlight glinting on distant satellites that are otherwise too faint to be seen with the naked eye.

INSTRUMENTS AND OBSERVING

BINOCULARS are the ideal instrument with which to begin observing. They are portable, have a wide field of view, are relatively inexpensive and will still be of use even if you later progress to a telescope. Surprisingly enough, there are some branches of observation for which binoculars are preferable to telescopes.

Almost any pair of binoculars can be pressed into service for looking at the sky, but there are certain points to consider when buying binoculars specifically for astronomy. Most importantly, do not be tempted by claims of high magnification. Unless you have a tripod on which to support the binoculars, a magnification of 10 times will be the most that you can use satisfactorily without the image becoming hopelessly unsteady due to movement of your hands and arms – each tremor is magnified as much as the image. Remember also that larger binoculars are heavier and more tiring to hold up.

Binoculars are really nothing more than a pair of telescopes joined together. It is more restful to view with both eyes simultaneously than to peer one-eyed through a telescope. Binoculars are more compact than telescopes because the path of the light is folded by prisms within the casing. In cheaper binoculars these prisms are not always securely fastened and can shake loose if subjected to rough treatment.

Most binoculars have a central focusing wheel that allows you to home in rapidly on objects of interest. One eyepiece is always adjustable to allow for inevitable differences in vision between your two eyes. Avoid binoculars that can be focused only by adjusting both eyepieces individually; these are cumbersome to use. Zoom binoculars with an adjustable range of magnification are available, but for best optical quality it is advisable to choose binoculars of fixed magnification.

Binoculars are specified by designations such as 7×50 or 8×40. The first figure is the magnification, and the second figure is the *aperture* – the diameter of the front lenses expressed in millimetres. A magnification of 8 times, for example, means that the binoculars make the object appear 8 times larger than to the naked eye and hence 8 times closer (although stars will still appear only as points of light). Binoculars usually have a magnification between 6 and 10 times, and aperture from 30 to 50 mm, but there are larger and more powerful instruments (for example, 20×70) which can be used on tripods. However, a tripod defeats one of the advantages of binoculars – their portability.

There is a double price to be paid for magnification. Firstly, when an image is enlarged it becomes dimmer because the light is spread out over a larger area. So the greater the magnification, the more light is required to maintain an acceptably bright image. That can be obtained only with a larger collecting area, i.e. a wider aperture. Secondly, as the magnification applied to a particular instrument goes up so its field of view shrinks.

To get a satisfactory field of view and a bright image, you should look for a ratio of at least 1:5 between the magnification of the binoculars and the aperture of the front

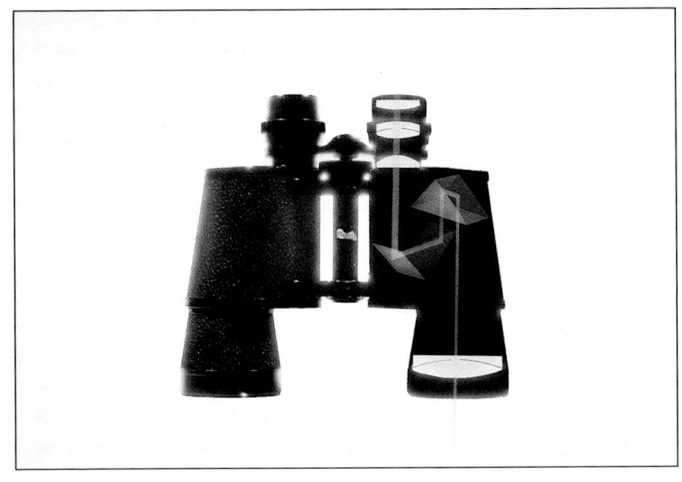

In binoculars the light path is folded by prisms.

lenses. If the ratio between the magnification and aperture is small (for example, 8×20) the field of view will be narrow and the image will be dim. Specifications such as 6×30, 8×40, 7×50 and 10×50 are all good for astronomy. Other things being equal, larger apertures are preferable because they collect more light and thus show fainter objects. Good binoculars will have a field of view of 6° or more; the field of view is often marked on the instrument. For comparison, the typical field of view of a telescope with a low-power eyepiece is about 1°.

Stargazing is the severest test that can be applied to any optical instrument, since the slightest imperfections show up when you are trying to focus stars into crisp points of light. A pair of binoculars that appears to perform perfectly well on daytime scenes can give mediocre views of the night sky. Hence it is impossible to be sure how good binoculars are until you have tried them out under operating conditions. It is worth buying the best pair that you can afford.

Incidentally, if you are short sighted or long sighted you will not need to wear your spectacles when observing since you can adjust the focus of the binoculars to compensate. However, if you suffer from any other defect, such as astigmatism (in which points of light appear as streaks), you will need to wear your glasses. Unfortunately, this increases the separation between your eye and the eyepiece and cuts down the field of view; it is best to wear contact lenses.

What are binoculars used for? They are ideal to help

beginners find their way around the constellations, particularly when viewing from towns where the fainter naked-eye stars are swamped by pollution and artificial lighting. Star colours are more noticeable through binoculars – look, for example, at Arcturus and Betelgeuse and see how much redder they look through binoculars than to the naked eye. (The colour of a star indicates its temperature – see page 102.) Wide double stars, such as ε Lyrae, are more easily separated through binoculars than with the naked eye. Observers of the brighter variable stars use binoculars when comparing their brightness against neighbouring stars, and some amateurs make binocular searches for the erupting variable stars known as novae.

The wide field of view of binoculars is particularly suited to sweeping over the star fields of the Milky Way and for scanning star clusters. Indeed some star clusters, such as the Pleiades, the Hyades and Praesepe, are so large that only binoculars can show them in their entirety. Binoculars are ideal for picking out extended, diffuse objects such as nebulae (gas clouds) and certain galaxies such as M33 in Triangulum. For the same reason binoculars are also ideal for studying comets, which are fuzzy and indistinct.

Closer to home, binoculars are ideal for following artificial satellites as they cross the sky. Held steadily, binoculars will reveal Jupiter's four main moons – tiny starlike points that change position from night to night as they orbit the planet. Binoculars will help to find Mercury, which is always low down in the twilight, and will show the crescent of Venus when that planet is close to us. They will reveal the distant planets Uranus and Neptune, and the brightest asteroids.

Our own Moon presents a stunning sight with even the lowest magnification, constantly changing appearance as it passes through its cycle of phases, and lunar eclipses are well seen through binoculars. *Never look at the Sun through any form of optical instrument, though – it is so bright that it will blind you.* If you wish to observe the Sun, hold the binoculars (or preferably mount them) so that you can project the Sun's image onto a piece of white card.

TELESCOPES come in two main types: *refractors*, which use lenses to collect and focus light, and *reflectors*, which collect light with a large mirror. Combination designs, called *catadioptric* telescopes, use lenses as well as mirrors but are best thought of as modified reflectors. The smallest

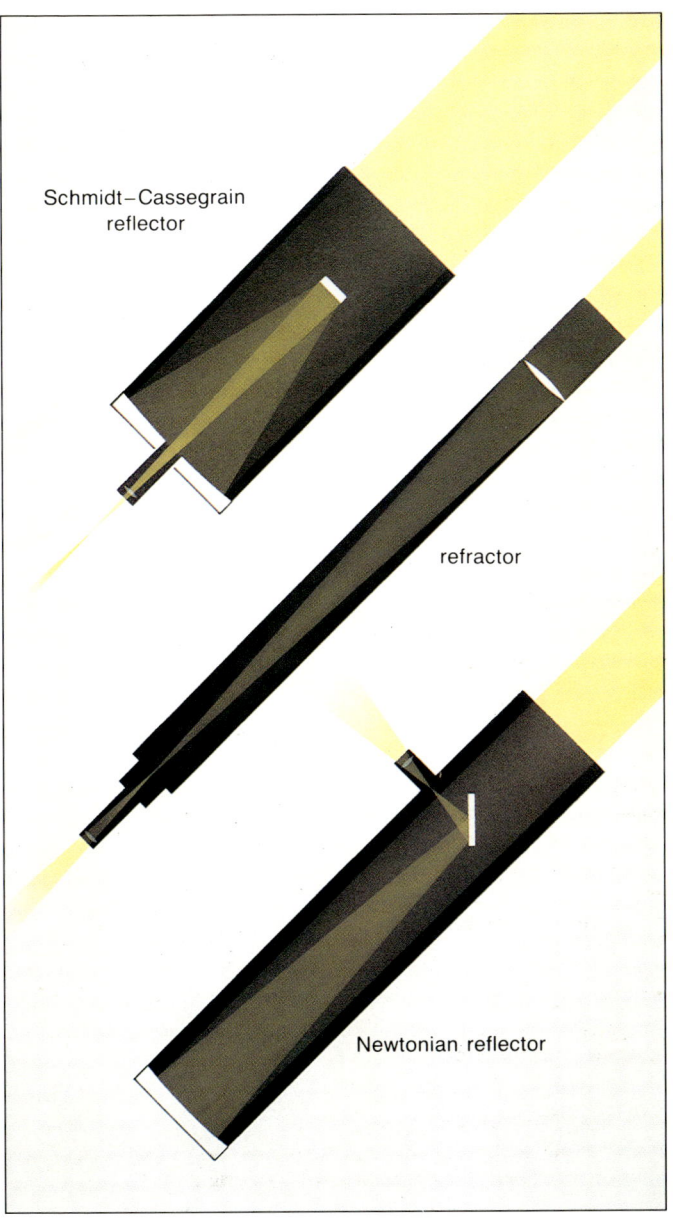

Schmidt–Cassegrain reflector

refractor

Newtonian reflector

Types of telescope.

telescopes, below about 100 mm (4 inch) aperture, are refractors while all the largest telescopes, above 1 metre (40 inch) aperture, are reflectors, for reasons to do with their design and manufacture. Intermediate-sized telescopes can be either refractors or reflectors, but reflectors tend to be the most popular with amateurs because, for a given aperture, they are significantly cheaper than refractors. Note that a telescope is always referred to by its aperture. A telescope's aperture governs its performance, more so than any other factor, and therefore the best advice when choosing a telescope is to get the largest aperture you can afford.

Refracting telescopes were the first to be invented, early in the 17th century. A main lens, called the *object glass*, collects light and focuses it down a tube where it is magnified by an eyepiece. Lenses are made of two or more pieces of glass to ensure that light of all colours is brought to the same focus. Cheap refractors (and cheap binoculars) do not adequately focus all colours to the same point, with the result that images are surrounded by yellow and blue fringes, a defect known as *chromatic aberration*. The cause of chromatic aberration is that light of different wavelengths is refracted through the lens by different amounts. The most extreme example of this effect is when white light is split by a prism into a rainbow-coloured spectrum.

Reflecting telescopes do not suffer from chromatic aberration, since all the light bounces off the front surface of a mirror and does not undergo refraction. The first reflector was made in 1668 by Isaac Newton specifically to overcome the problem of chromatic aberration that plagued early refractors. In his design, still known as the *Newtonian* reflector, light is focused by a concave main mirror onto a small flat mirror or prism which deflects the light-beam to the side of the tube, where an eyepiece is placed. The Newtonian reflector remains the most popular design of telescope for amateur astronomers. An extreme type of Newtonian is the *rich-field telescope*, which uses a mirror of very short focal length and a low-power eyepiece to give a field of view of 3°, not as good as binoculars but still convenient for general stargazing.

An alternative design, called the *Cassegrain* after the Frenchman who invented it, uses a convex secondary mirror to reflect the light back through a hole in the main mirror, where an eyepiece or other instrument is placed. Odd though this system may at first seem, it is widely used in large professional telescopes and is also becoming increasingly popular among amateurs in the so-called *Schmidt–Cassegrain*

CHOOSING A TELESCOPE depends on the type of observing you intend to do, as well as your budget. Many amateurs start with small refractors of 50 to 60 mm (2 to 2.4 inches) aperture. These are sufficient to show details on the Moon and planets, to find the brighter galaxies and planetary nebulae, and to separate many well-known double stars. They are particularly suitable for projecting the image of the Sun, which is so bright that large apertures are not needed. In fact, this is where refractors are preferable to reflectors since the Sun's concentrated heat and light can damage a reflector's secondary mirror.

For larger sizes, the cost advantage of the Newtonian reflector usually outweighs all other considerations. Newtonians in the 150 mm (6 inch) to 250 mm (10 inch) range are common in amateur astronomy. Their only real drawback is that periodically the mirror needs to have its reflective coating renewed, although each coating should last for many years.

However, if you intend to specialize in areas such as planetary observation or close double stars, where resolution of fine detail is critical, then the slightly better definition available with a refractor may be worth the additional expense. If compactness and portability are the governing factors (for example, if you live in a flat), then catadioptric telescopes have the advantage. Catadioptrics are particularly popular for astrophotography.

variant, where the secondary mirror is attached to the centre of a thin lens called a correcting plate at the top of the telescope tube. This increases the distortion-free field of view of the telescope. The design was introduced by the Estonian astronomer Bernhard Schmidt in 1930, and is now used for wide-angle photographic telescopes. Such catadioptric telescopes are expensive, though.

Inevitably, the secondary mirror of a reflector acts as an obstruction that cuts down the amount of light reaching the main mirror and can slightly degrade the image. In the smallest telescopes this disadvantage is serious, which is why reflectors are not made with apertures of less than about 100 mm (4 inches). For larger apertures the effects of the secondary obstruction become proportionally much less, and the advantages of the reflector heavily outweigh its disadvantages. For one thing, reflectors are much more compact than refractors of the same aperture since mirrors can be made with shorter focal lengths than lenses. Another advantage of reflectors is purely economic: large mirrors are much cheaper to make than large lenses.

A telescope mirror is supported at numerous points on its

rear surface so that it does not go out of shape, and the trend nowadays is to very thin, lightweight mirrors. A telescope lens, though, is held only around its rim, so the glass must be thick enough to support its own weight. Thicker glass absorbs more light, so there is a practical limit to the size of a telescope lens. The largest refracting telescope, at Yerkes Observatory in Wisconsin, USA, has a lens 1.02 metres (40 inches) in diameter; it was built at the end of the last century and is unlikely to be exceeded in size. Reflectors, on the other hand, have been made with mirrors up to 6 metres (236 inches) in diameter, and even larger sizes are now possible with mirrors made in segments.

Low-expansion forms of glass such as Pyrex, and more advanced types with names such as Zerodur, are used for telescope mirrors. Unlike household mirrors, where the reflective coating is on the rear, telescope mirrors are coated on their front surface so the light does not travel through the glass at all. A telescope mirror's reflective coating is made of aluminium, deposited as a very thin film, and reflects nearly 90% of the incident light. Centimetre for centimetre a refractor should always perform slightly better than a reflector because there is no central obstruction, but since a larger reflector can be had for the price of a refractor of smaller aperture this advantage is irrelevant.

TELESCOPE PERFORMANCE depends on the aperture and the focal length of the objective lens or main mirror. Aperture is the factor that determines the faintest objects and finest detail visible, while focal length affects the size of the image and field of view.

There can be no hard-and-fast rules about how much you will see with a given telescope, since so much depends on its optical quality, the local atmospheric conditions and the observer's eyesight. However, *in theory*, the magnitude of the faintest stars that you can expect to see with a telescope of aperture D (in millimetres) is given by the formula

$$m = 2.7 + 5 \log D$$

but in practice the real performance may fall well short of this. For example, if you are observing from a town the faintest stars visible in a 100 mm (4 inch) reflector are likely to be around ninth magnitude.

Similar reservations apply to any attempt to specify the closest pair of stars that can be separated, a quantity known as the *resolution* or *resolving power* of the telescope. The theoretical limit on the resolution of an aperture D (in millimetres) is given by the formula

$116/D$

and the result is expressed in arc seconds. The table shows the limiting magnitudes and resolutions for a range of apertures calculated from the formulae.

Aperture, *D*	Limiting magnitude	Resolution (arcsec)
50 mm (2 inches)	11.2	2.3
60 mm (2.4 inches)	11.6	1.9
75 mm (3 inches)	12.1	1.5
100 mm (4 inches)	12.7	1.2
150 mm (6 inches)	13.6	0.8
200 mm (8 inches)	14.2	0.6
250 mm (10 inches)	14.7	0.5

The other factor affecting the performance of a telescope, its focal length, is the distance between the lens or mirror and the point at which it brings starlight to a focus. Usually, though, instead of the focal length the *focal ratio* or *f*-ratio is given since it is a more convenient way of comparing different telescopes. Focal ratio is the focal length of a lens or mirror divided by its aperture. For example, a telescope of 150 mm aperture and 900 mm focal length has an *f*-ratio of 6, written *f*/6.

Traditionally, refractors have had focal ratios in the range *f*/10 to *f*/15 since in the past it was difficult to make a good lens with short focal length. Nowadays, though, with the use of special fluorite glass for objective lenses, refractors can be *f*/7 or even shorter. Typical Newtonian reflectors have *f*-ratios of around *f*/8 to *f*/6 (or *f*/4.5 for rich-field versions). Catadioptric telescopes usually have *f*-ratios around *f*/10, but again modern optical technology is allowing the production of previously unattainable short *f*-ratios similar to those of Newtonians.

Long focal ratios, as in traditional refractors and most catadioptrics, mean a large image size and hence small fields of view (although the type of eyepiece used also affects the field of view – see page 32). For detailed planetary work a large focal ratio is ideal, but for wide-field stargazing smaller *f*-ratios are preferable. The intermediate focal ratios of Newtonian reflectors make them good all-round instruments. Telescopes usually come fitted with a small, low-power telescope called a *finder* with a wide field of view which is indispensable for locating objects.

EYEPIECES AND MAGNIFICATION. Astronomical telescopes have interchangeable eyepieces that offer different magnifications. For wide-field views of extended objects such as star fields you would use a low power, while for close-ups of the Moon and planets you would switch to a high power.

How much a telescope magnifies depends on two factors: its focal length and the focal length of the eyepiece being used with it. For a particular telescope, an eyepiece of short focal length will give a higher magnification than an eyepiece of long focal length, but the same two eyepieces would give different magnifications in other telescopes.

MAGNIFICATION is calculated by dividing the focal length of the telescope by the focal length of the eyepiece. For example, a telescope of 900 mm focal length and an eyepiece of 30 mm focal length will magnify 30 times. The same eyepiece, applied to a telescope of 1200 mm focal length, will magnify 40 times. An eyepiece of half the focal length, 15 mm, will double the magnification in each case, i.e. to 60 and 80 times. Magnification is often expressed in the form ' × 100' (100 times).

It is important to realize that there is a practical limit to the amount of magnification you can apply to a given telescope. Too high a magnification will produce an image that is faint and indistinct, whereas crisper views are obtained with more modest powers. The explanation goes back to the matter of aperture: large apertures collect more light and allow higher magnification.

In practice, the highest magnification that can effectively be used on a telescope is about twice its aperture in millimetres. That means 150 times on a telescope of 75 mm (3 inch) aperture, 200 times with 100 mm (4 inch) aperture, and so on. But no matter how large your telescope, you will rarely be able to use a power of more than about 400 times because of the unsteadiness of the Earth's atmosphere (see page 35). A collection of three eyepieces offering a range of powers is usually sufficient. A poor eyepiece will ruin the performance of even the finest telescope, so it is false economy to cut corners when choosing eyepieces.

Somewhat confusingly, different designs of eyepiece have different fields of view, even though they may be of the same focal length. A field of view is sometimes stated on an eyepiece or in its advertising, and is usually between about 30° and 80°. This is the *apparent field of view*. It is not the field of view you will actually see, though, for the final result also depends on the focal length of the telescope itself. To find the true field of view for your telescope, divide the apparent field of the eyepiece by the magnification it produces. For example, an eyepiece with 50° apparent field of view that magnifies 40 times shows you a field of 1¼°.

An easy way to measure your telescope's field of view is to time a star near the celestial equator as it crosses the field. The Earth rotates through 0.25 minutes of arc in 1 second of time, so if the star takes 60 seconds to cross, the field is 60 × 0.25 = 15 arc minutes wide.

Another feature of eyepieces frequently mentioned in advertising is their *eye relief*. This is, in effect, how close you have to place your eye to the eyepiece in order to see the complete field of view. Good eye relief means that you do not need to cram your eye right up to the eyepiece, so those who wear glasses should choose eyepieces with good eye relief.

TELESCOPE MOUNTINGS are as important as the optics, since a telescope is of little use if it cannot be kept steadily aimed at the object of interest. The most basic type of mounting, used for simple refractors and some small reflectors, is called an *altazimuth*. The telescope is simply mounted on a tripod and is free to pan around in the horizontal plane (azimuth) and pivot up and down in the vertical plane (altitude). This is the type of mounting found on coin-in-the-slot telescopes at scenic spots.

A modified form of altazimuth mounting has become popular in recent years for short-focus reflectors. It is named the *Dobsonian* mount after its inventor John Dobson, an American amateur. The secret of the Dobsonian is that all the weight is concentrated at the mirror end of the tube; the telescope sits on a simple wooden frame that turns on Teflon pads. For cheapness and portability nothing beats the Dobsonian, as long as low-power observing is all that is required.

Altazimuth mountings have the disadvantage that they must be continually adjusted in both axes to compensate for the motion of the sky as the Earth rotates (1° in right ascension every 4 minutes). Objects are much simpler to track with an *equatorial* mount. Here, one axis (called the polar axis) is aligned so that it points directly at the north celestial pole. The sky appears to rotate around the celestial pole, so, once the telescope is pointed at an object, simply turning the polar axis will keep the object in view, a job that can be done by an electric motor. The other axis of the mounting is called the declination axis, since it allows the telescope to move up and down in declination.

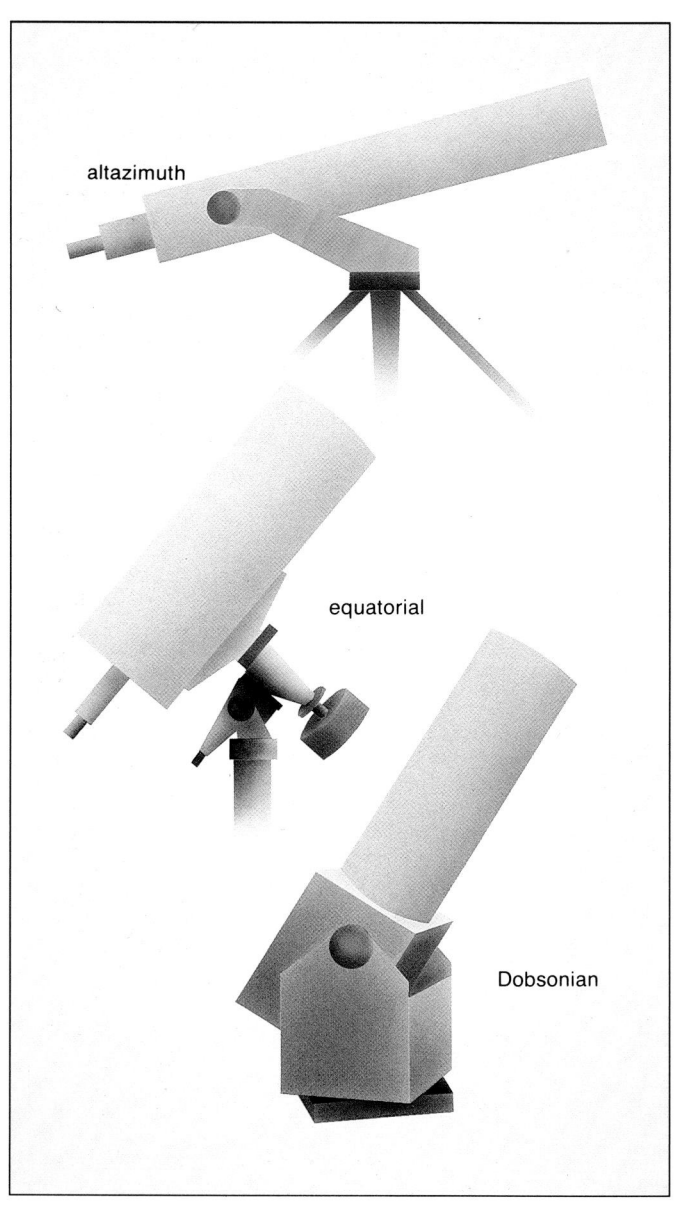

altazimuth

equatorial

Dobsonian

Telescope mountings.

There are several varieties of equatorial mounting, the most usual being the German type in which the telescope is counterbalanced by weights on the declination axis. A fork-type mounting is widely used for catadioptric telescopes since it is particularly suited to their short tubes. The arms of the fork point towards the north celestial pole, and the telescope tube is pivoted between them on the declination axis.

OBSERVING. 'It's upside down!' is the usual reaction of someone looking through an astronomical telescope for the first time. To turn the image the right way up would require an extra lens in the eyepiece. This is provided in binoculars, since they are designed for everyday use. But each additional piece of glass reduces the amount of light reaching the observer's eye (and adds to the cost). It does not matter to an astronomer which way up the image is, since there is no 'up' or 'down' in space, but any loss of light *would* matter, so the extra lens is left out of astronomical telescopes and the image remains inverted.

Even a brief look through a telescope reveals the astronomer's worst enemy – unsteadiness of the atmosphere, which causes the image to shimmer as though seen through a heat haze. Astronomers use the term *seeing* – when the image is steady the seeing is said to be good. At times of bad seeing the image 'boils' so much that the resolution of the telescope is seriously impaired, and the images of close double stars merge. The effects of seeing become more noticeable under higher powers, and the seeing is always worse close to the horizon.

Seeing should not be confused with *transparency*, which is a measure of the clarity of the atmosphere. The better the transparency, the fainter the objects that can be seen. Good transparency and good seeing do not necessarily go hand in hand: sometimes, when a weather front has passed, for example, transparency is excellent but seeing is atrocious because of the mixing of warm and cold air. On nights like this, the stars twinkle brilliantly. Such nights are not good for splitting close double stars or observing fine detail on the planets, but they are ideal for observing extended objects such as nebulae, galaxies and comets, which are not affected by bad seeing. Conversely, when transparency is poor, as on misty nights, seeing can be excellent because the air is calm. Transparency shares one characteristic with seeing: it is always poorest close to the horizon, so to get the brightest, steadiest image objects should be observed when they are as high in the sky as possible.

If you go out into the night from a brightly lit room you will not, at first, be able to see very much. As your eyes become used to the dark you will see progressively fainter objects; this process is known as *dark-adaptation*. Before looking at faint or diffuse objects such as nebulae, galaxies or comets, you should allow at least ten minutes for your eyes to become suitably dark-adapted. In fact, dark-adaptation can continue to improve slowly for at least half an hour, but it is immediately destroyed on exposure to light again. To preserve your dark-adaptation while reading a star map or writing down observations, it is wise to use a torch (flashlight) with a bulb that has been painted red or has a red filter over it, since red light has the least effect on dark-adapted eyes. In an observatory you can use the kind of red light used in a photographic darkroom.

One useful trick for seeing very elusive objects is *averted vision* – looking slightly to one side of the object of interest. This has the effect of allowing its light to fall on the outer part of the retina of the eye, which is more sensitive than the central part. Another trick is to tap the telescope gently so that the image vibrates. Surprisingly, this can make elusive objects easier to see.

It is important to keep an observing book in which to write down the details of each observation and make sketches. Firstly, always note the date and time of each observation. Astronomers record dates in the form year/month/day and then, if necessary, decimals of a day. Times of astronomical events are usually given in Universal Time (UT), which is the same as GMT. If you are in a different time zone from Greenwich and prefer to use your local time (e.g. Eastern Standard Time for the eastern seaboard of the USA) then be sure to note which time system you are using. Also remember to allow for the hour's difference when summer time (daylight saving) is in operation. Note the instrument and magnification used, and the sky conditions.

Finally, one important but often overlooked part of an astronomer's equipment is warm clothing. Even on a seemingly mild night a tracksuit and training shoes make a practical outfit, while in the depths of winter you will need to dress like a mountaineer, with a quilted jacket, thick trousers and boots.

An amateur astronomer observes the crescent Moon.

THE SUN AND THE MOON

THE SUN. Beginners often regard astronomy as an exclusively nocturnal activity, and so overlook the most obvious celestial body of all – the Sun. The Sun is a typical star, a ball of glowing gas that gives out the heat and light that keep us alive here on Earth, and it is the only star whose surface we can see in close-up. Since we are totally reliant on it for our existence, the Sun is of particular interest to astronomers. It might be thought that the Sun is featureless, but that is not so. Dark patches called *sunspots* mark its surface, each lasting from a few days to several weeks, rising and falling in number over a cycle lasting approximately 11 years. It is the counting and tracking of sunspots that is the main occupation of amateur solar observers.

Most astronomical objects cause a problem for observers because they are so faint, but the problem with the Sun is exactly the opposite. It is so bright that it must be treated with extreme caution. *You should never, ever, look directly at the Sun through any form of optical aid.* Even the briefest glimpse can focus enough heat and light onto your eye to cause permanent blindness.

The only exception to that rule is if you are using a trustworthy filter over the front of the telescope (*not* over the eyepiece). Solar filters consist of sheets of thin plastic or glass with a metal coating that blocks all but a tiny fraction of the incoming sunlight, allowing you to view in safety. Depending on the nature of the coating, the resulting image will be orange or blue in colour. However, solar filters are prone to scratching or cracking – with potentially disastrous results – and they are costly. What's more, they are unnecessary since the Sun can be safely observed by projecting its image onto a white surface. This technique has the added advantage that several people can view the Sun simultaneously, and it is strongly recommended to all beginners.

SUN DATA BOX

Diameter: 1,392,500 km
Mass: 333,000 × Earth
Mean density: 1.41 × water
Volume: 1.3 million × Earth
Escape velocity: 617.3 km/sec
Mean sidereal rotation period:
  25.38 days
Mean synodic rotation period:
  27.28 days
Apparent magnitude: −26.8
Absolute magnitude: 4.8
Mean distance from Earth:
  149,600,000 km

Since the Sun is so bright, only a small refractor is needed to observe it. If you have a large reflector it is advisable to stop down the aperture to about 100 mm to prevent excessive heat from entering the optics. Be very careful while aiming the telescope at the Sun. Squint along the tube (do *not* use the finder!), keeping the covers on the lens (or mirror) and the finder so that no light can accidentally enter your eyes. Alternatively, line up the telescope on its own shadow. Then remove the lens cap and focus the Sun's image onto a white card (or even a smooth, light-coloured wall). Use a low-power eyepiece to see the whole disk.

Some people attach a solar observing box made out of cardboard or lightweight wood to the eyepiece end of the telescope, while others make do with a large piece of card fitted over the telescope tube to cast a shadow onto the

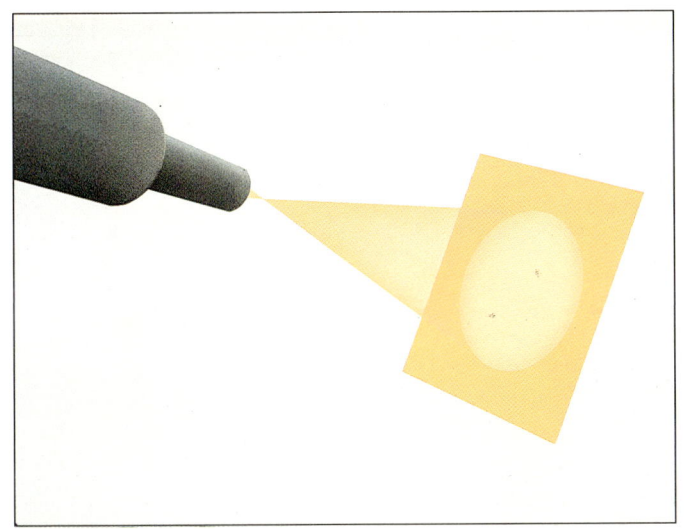

The safest way to observe the Sun is by projecting its image.

screen. If you are using a Newtonian reflector, a projection screen will be easier to shield since it will be facing at right angles to the direct sunlight.

As you look at the projected image of the Sun, you will see that the edges of the disk are somewhat darker than the centre. This *limb darkening* occurs because the Sun is gaseous, not solid, and its outer layers are more tenuous than its denser interior. Against the limb-darkened regions can be seen brighter patches known as *faculae*, which are hotter areas

The Sun's disk, showing limb darkening and some sunspots.

of gas on the Sun's surface that appear before the formation of sunspots and remain after their subsequent decay.

Sunspots are cooler areas of gas on the Sun's surface where strong magnetic fields affect the outflow of heat from inside the Sun. Each spot has a dark, central area called the *umbra*, surrounded by a lighter *penumbra*. The Sun's visible surface, termed the *photosphere*, has a temperature of about 5500°C. A sunspot's penumbra is a few hundred degrees cooler than this, while the umbra is over 1000°C cooler than the photosphere. Sunspots often appear in pairs that act like the north and south poles of a horseshoe magnet.

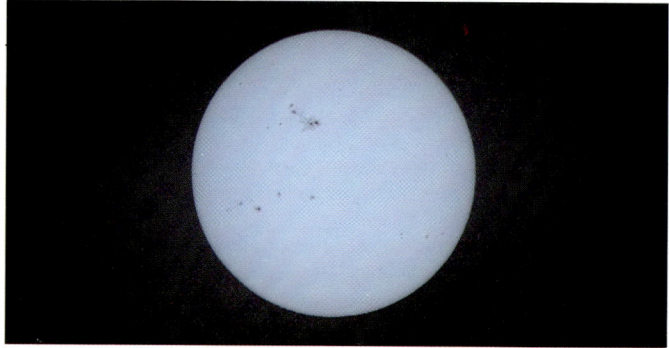

Large sunspot group of 1989 June 13, photographed through an 800mm telephoto lens at f/32, 1/1000 second exposure on ISO 100 film, through a Mylar filter (which gives the bluish cast).

Watching sunspots develop and decay is one of the fascinations of solar observing; the Sun is never the same two days running. All the spots that you will see are larger than the Earth, and complex groups can stretch for up to 100,000 km. The largest spot groups are visible to the naked eye, either through a filter or when the Sun is dimmed by the atmosphere just before setting.

You may wish to make a sketch of the entire disk of the Sun, and perhaps add a close-up of any large groups. A circle of 150 mm (6 inches) diameter is standard for solar disk drawings. It helps to divide the circle into a reference grid of 1 cm squares so that the positions of sunspots can be accurately copied onto a chart. Look carefully at sunspots near the limb. Sometimes the penumbra is wider on the side closest to the limb, giving the spot a dished appearance. This is known as the *Wilson effect*, after the Scottish astronomer who first described it over 200 years ago. Paradoxically, a few spots exhibit the opposite effect and appear slightly dome-shaped.

By comparing the position of the spots from day to day you will be able to measure the rotation of the Sun. Being gaseous, the Sun does not rotate all at the same rate. At the equator it rotates in about 25 days, slowing to 36 days near the poles. These rotation periods are relative to a fixed external point, and are termed the *sidereal* periods. Since the Earth is not fixed but is moving in orbit around the Sun, the time the Sun takes to rotate once with respect to the Earth (known as the *synodic* period) is about two days longer. As seen from the Earth, therefore, a spot near the equator would take just over 27 days to go once around the Sun, but not many of them last that long.

Sunspots are confined between about latitudes 40°N and 40°S. Measuring latitudes on the Sun is not straightforward. You can find the east–west line in the sky by allowing a spot to drift across your projection screen, but the tilt of the Sun's axis from the vertical must also be taken into account; its values for different dates during the year are given in annual handbooks of astronomy. Note that in the projected disk of the Sun, north is at the top.

Sunspots and faculae are both classed as *active areas*, and one task of amateur astronomers is to count the number of active areas visible each day to get an indicator of changing solar activity. At times of high activity perhaps two dozen small spots and several large groups may be visible at a time, whereas at minimum the Sun's disk can be blank for days on end. Solar activity rises and falls in a cycle lasting

SUNSET. The Earth's atmosphere reddens and dims the setting Sun and distorts it into unusual shapes. Sometimes, particularly when the Sun is setting over the sea, the last remaining segment of its disk turns green for a second or two. This is known as the *green flash*, and is caused by atmospheric refraction. The green flash is best seen when the atmosphere is clear and there is relatively little reddening of the Sun.

approximately 11 years. The last maximum was in 1990, to be followed by a minimum in the mid-1990s.

There are other forms of activity on the Sun, such as *prominences*, which come to a peak along with sunspots. Prominences are jets or loops of gas extending from the photosphere, but they are not visible without special filters except during a total eclipse, when the brilliant photosphere is obscured. Rarely, a brilliant eruption called a *flare* may be seen near a sunspot. Flares eject streams of atomic particles into space which, if they encounter the Earth, cause glows in the upper atmosphere known as *aurorae*.

AURORAE, popularly known as the northern (or southern) lights, are the product of atomic particles from the Sun being channelled by the Earth's magnetic field towards the north and south magnetic poles. Cascading down onto the upper atmosphere, the atomic particles cause colourful glows at altitudes of 100 km and above. Aurorae are commonly seen in far northern regions such as Alaska, Canada, Scotland and Scandinavia. Only occasionally do they extend farther south, notably following a solar flare. One of the most extensive auroral displays of the century in March 1989 was seen as far south as the Caribbean, but this is exceptional. Major aurorae are accompanied by interference to radio communications and disturbances to power lines, resulting from the disruption of the Earth's magnetic field.

The colourful glow of an aurora on 1989 April 26, photographed with a 50mm lens at f/1.8, exposure 20 seconds on ISO 64 film.

A simple aurora consists of a glow on the northern horizon (the word aurora means 'dawn'). Over the next hour or so the display can develop into an iridescent arch with bright rays and streamers extending vertically through it, rising high in the sky before collapsing again. Some aurorae take the form of folded curtains, gently rippling as though blown by a breeze. At their most spectacular, aurorae can resemble laser light shows and can last all night. Red and green are the usual colours, caused by oxygen atoms in the atmosphere glowing like the gas in a fluorescent tube. In towns, though, aurorae are difficult to see because they are swamped by artificial lighting.

Since aurorae are closely linked with activity on the Sun they are most common around times of sunspot maximum. Look out for an aurora a couple of nights after a large, complex sunspot group is near the centre of the Sun's disk – only when the spots are near the centre will the Earth be in the firing line of any flares, and the atomic particles from the Sun take a day or two to reach the Earth. Although aurorae can never be predicted, it is possible that 27 days later, when the Sun has completed one full turn, there may be a repeat performance if the same area is still active.

THE MOON is our nearest celestial neighbour and is the most spectacular sight of all for users of small instruments. Even the naked eye can detect some coarse detail on the Moon, notably the dark patches that make up the familiar 'man in the Moon' pattern (alternatively interpreted by some people as a rabbit or a crab). Turn a pair of binoculars or a small telescope on the Moon and you will be able to see the main features depicted on the maps on pages 52 to 67.

Immediately noticeable is the smoothness of the dark lowlands in contrast to the bright, heavily cratered uplands. Take a look along the *terminator*, the line dividing day and night on the Moon. Here you will see lunar features thrown into sharp relief by the low angle of lighting. There is no atmosphere to soften the shadows. Under a higher angle of illumination the surface relief blends into the

---

MOON DATA BOX

Diameter: 3476 km
Mass: 0.0123 × Earth
Mean density: 3.34 × water
Volume: 0.02 × Earth
Escape velocity: 2.37 km/sec
Mean sidereal rotation period: 27.32 days
Apparent magnitude (full): − 12.7
Mean distance from Earth: 384,400 km

background, but the contrast between dark and light areas becomes more prominent.

Craters of all sizes abound on the Moon, although the larger ones are mostly restricted to the highlands. Look for example at magnificent Clavius in the southern hemisphere, 225 km in diameter and with a chain of smaller craters arcing across its floor. All but a few of the craters visible through small telescopes were formed long ago when asteroids, comets and meteorites smashed into the lunar surface at high speed; the exceptions are some small craters that may have a volcanic origin. This bombardment is still going on, but the meteorites hitting the Moon today are much smaller and the craters they produce are too small to be seen through a telescope. The craters are named after prominent scientists of the past.

Some craters are surrounded by bright rays, seen most prominently under high illumination, consisting of material thrown out in the impact. Examples are Copernicus, 90 km in diameter, and Aristarchus, 40 km, which is the brightest area on the Moon. Perhaps the most magnificent rayed crater of all is Tycho, in the southern hemisphere near Clavius. The crater itself is 84 km in diameter but its rays stretch for a thousand kilometres. These rayed craters are believed to be among the youngest on the Moon, although 'young' in this context means an age of a few hundred million years, against the 3 billion years or more of the rest of the surface.

Early observers termed the Moon's dark areas *maria* (singular *mare*), Latin for seas, which is what they were

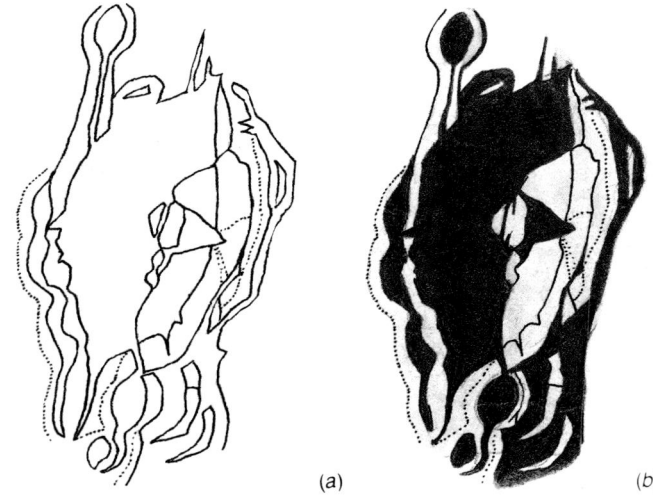

(a)   (b)

thought to be. We have long known that there is no air or water on the Moon but the term has persisted, so that we have poetically named areas such as Mare Serenitatis, the Sea of Serenity, and Oceanus Procellarum, the Ocean of Storms. The mare areas were evidently formed by the largest impacts of all. They were subsequently flooded, not with water but with lava, so perhaps the old names are not so inappropriate after all.

The term 'crater' is somewhat misleading since it conjures up the vision of a deep, circular bowl. In fact the largest craters are better regarded as walled plains, and many of them are far from circular. For example, the 150 km diameter Ptolemaeus near the Moon's centre is detectably hexagonal in outline, as is the 120 km Purbach further to the south. Presumably, these distorted shapes are due to stresses in the Moon's crust. Many large craters have terraced walls, apparently caused by slumping, and a central mountain massif, prominent examples being Theophilus, Arzachel and the rayed crater Copernicus.

There is a near-continuous range of sizes from the largest craters to the smallest maria. Most crater-like of the maria in appearance is Mare Crisium, a basin approximately 500 km in diameter, twice the width of the largest craters. Some craters have been flooded with dark lava so that they have mare-like floors, notably 100 km Plato on the northern shore of Mare Imbrium (Sea of Rains). Near Plato is a huge half-crater called Sinus Iridum (Bay of Rainbows), 260 km across, which

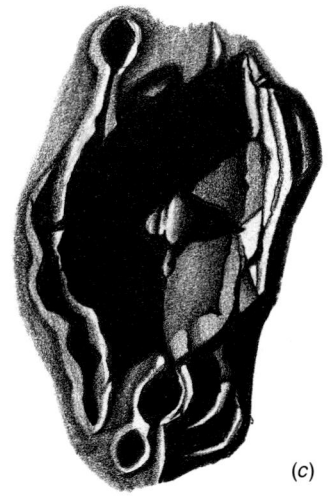

How to draw lunar features: Begin by sketching an outline of the features with a hard pencil (a). Then fill in the shadows with a soft pencil, shading more lightly to indicate differences in surface tone, and leaving the brightest areas white (b). Finally, add the finest detail visible during moments of steady seeing (c). The crater depicted is Piccolomini, as seen through a 60 mm refractor, × 100.

(c)

effectively bridges the gap between craters and maria. It opens into Mare Imbrium and only a few low ridges, visible under oblique sunlight, remain where its 'seaward' wall was destroyed by the advancing lava. The wall of Sinus Iridum that still stands forms an arcuate range called the Jura Mountains, rising 4500 m high. Sunrise on the Jura Mountains, 10 days after new Moon, is one of the most impressive sights for lunar observers.

The demolished wall of Sinus Iridum is a type of feature known as a *wrinkle ridge*. Many other such ridges can be found on the various maria, notably the Serpentine Ridge on Mare Serenitatis. Wrinkle ridges extend for hundreds of kilometres but are no more than a few hundred metres high, so they can be seen only under low angles of illumination. No one is entirely sure how wrinkle ridges formed, but there is probably more than one cause. Some may be places where the mare surface has shrunk; others are most likely the result of lava welling upwards through fissures; while still others, for example Sinus Iridum, seem to be the remains of crater walls that have been melted down. When the Moon is a crescent, about 5 days old, look for a ghostly feature on Mare Tranquillitatis called Lamont, near the crater Arago, which is composed entirely of wrinkle ridges.

Strong evidence of volcanic activity on the Moon is provided by *domes*, hummocky hills that appear to have been formed by thick lunar lava. The largest dome, or rather a grouping of them, is a strange excrescence on Oceanus Procellarum called Rümker, 55 km wide. Farther south on Oceanus Procellarum is a whole field of domes near the crater Marius.

Further evidence of internal activity on the Moon is provided by valleys of various kinds. One of the most obvious lunar valleys lies in the northern hemisphere, near Plato, and is called the Alpine Valley (Vallis Alpes in Latin) since it slices through the lunar Alps. The Alpine Valley seems to be a result of faults in the lunar crust, as do some much narrower features termed *rilles* which slice through mare surfaces and crater walls rather like wheel tracks, as for example around the rim of Mare Humorum (Sea of Moisture) and Fra Mauro and its neighbouring craters. Two prominent rilles lie near the centre of the the Moon's disk: the Ariadaeus rille and the Hyginus rille, the latter being dotted with small craters. Some lunar faults have produced shallow cliffs, as in the case of the Straight Wall (officially known as Rupes Recta), which stretches for 120 km down the eastern edge of the Mare Nubium (Sea of Clouds).

Even more fascinating are the *sinuous rilles* that snake

across the maria. Most prominent of this class is Schröter's Valley, which winds in a distorted W-shape for 200 km across Oceanus Procellarum, north of the crater Herodotus. Sinuous rilles look rather like dried-up river beds, and are thought to have been formed by streams of lava. The whole area of Schröter's Valley, in fact, is replete with examples of volcanic activity in the form of domes and sinuous rilles.

Finally, the most controversial aspect of lunar observation is the search for temporary changes such as hazy patches or coloured glows. Reports of such so-called *lunar transient phenomena* (LTPs) have been made by amateurs since the 1960s. If these reports are reliable, the LTPs must presumably be caused by releases of gas from inside the Moon or, occasionally, the impact of small meteorites. LTPs seem to cluster around the rims of certain maria, particularly in rilled regions. One favoured area is the crater Aristarchus, where in 1969 the crew of Apollo 11 saw an apparent brightening at the same time as observers on Earth reported an LTP. It is therefore always worth watching out for the unusual.

PHASES OF THE MOON. As the Moon orbits the Earth we see varying proportions of its sunlit hemisphere. When the Moon lies between us and the Sun, all the sunlit side is turned away from us and the Moon is said to be *new*. A day later the Moon emerges into the evening sky as a thin *crescent*. When the Moon is 'young', its night-time portion can sometimes be seen faintly illuminated by light reflected from

Earthshine (sunlight reflected from the Earth) illuminates the young crescent Moon. 10 second exposure with a 300mm lens at f/4.5 on a driven mount, ISO 400 film.

the Earth. This phenomenon is known as *earthshine*; it is sometimes called 'the old Moon in the new Moon's arms'.

After seven days we can see the Moon half-illuminated, at *first quarter*. Then the Moon fills out, passing through the *gibbous* phase, to become *full* approximately 15 days after new Moon, when it lies on the opposite side of the sky from the Sun and hence rises at about sunset. The phases then repeat themselves in reverse order, passing through *last quarter* on the way back to new Moon. The whole cycle is termed a *lunation* and lasts 29.5 days, an interval also known as the Moon's *synodic period*. The Moon rotates on its own axis in this same length of time, so that it keeps one face permanently turned towards the Earth. This is a result of the Earth's tidal forces acting on the Moon.

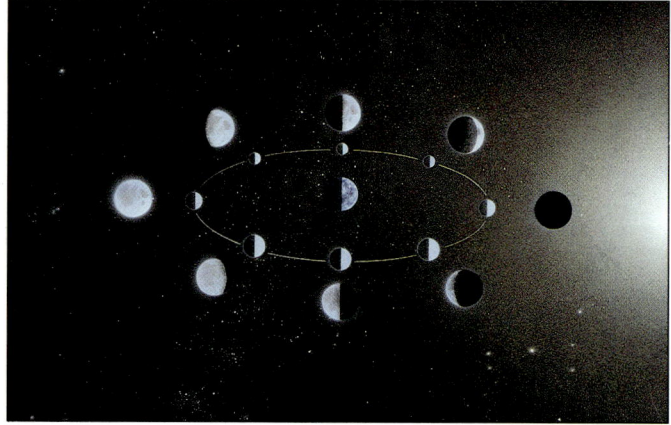

The Moon goes through a cycle of phases every month as it orbits the Earth and different proportions of its illuminated hemisphere are seen. Clockwise, from the right: new, crescent, first quarter, gibbous, full, gibbous, last quarter, crescent.

In practice, though, the Moon rocks slightly east–west and north–south so that at one time or another we can see up to 59% of its surface. This effect is known as *libration*, and is a result of both its elliptical orbit and its axial tilt. Because of libration, the features that lie on the terminator at a given phase can vary slightly from one lunation to the next.

LUNAR OCCULTATIONS are events in which the Moon moves in front of a star and obscures it. The precise timing of occultations by amateurs provides a simple yet effective way of tracking small changes in the orbit of the Moon that are caused by tidal interactions with the Earth. As a result of these tidal effects the Earth's rotation is gradually slowing and the Moon is gradually moving away from us, and occultation timings can help measure both these changes.

Predictions of occultations are published by national astronomical societies in their yearly handbooks and magazines. Since the Moon moves through its own diameter (half a degree) in an hour, it can occult a lot of stars in a night. However, predictions are usually restricted to stars of seventh magnitude or brighter, which gives an average of one or two a week. Sometimes the Moon will occult a star cluster such as the Pleiades, providing a feast of disappearances and reappearances during one night.

The technique is to start observing a few minutes before the occultation is due, in case the prediction is in error, with your finger on the button of a stopwatch (the stopwatches built into electronic wristwatches are highly convenient, but practise the technique beforehand). Start the stopwatch the instant the star disappears, and then stop it against a time signal such as the telephone time service. Deduct the time on the watch from the time given by the time service to obtain the time of the observation to the nearest tenth of a second. The exact latitude, longitude and altitude of the observing site must also be recorded; these can be found from a suitably detailed map.

Disappearances at the dark limb of the Moon are easier to record than those at the bright limb, since at the dark limb the star is not swamped by the Moon's glare. Reappearances of stars are far less easy to time accurately, since the observer is never quite sure exactly where the star will emerge, and can miss it. Disappearances (and reappearances) are instantaneous, since the Moon has no atmosphere to dim the star before it is finally obscured. If the star drops in brightness before finally disappearing, it must be a very close double. Occultation observations have revealed the existence of previously unknown close pairs, and have also shown up unexpected errors in the positions of certain stars.

Teams of amateurs travel long distances to observe *grazing occultations*, in which a star appears to brush the Moon's limb. These are of particular importance for they can reveal the Moon's position against the starry background with exceptional precision. Occasionally the Moon occults a planet rather than a star. These events are of less scientific importance, but they present an interesting spectacle.

continued on page 68

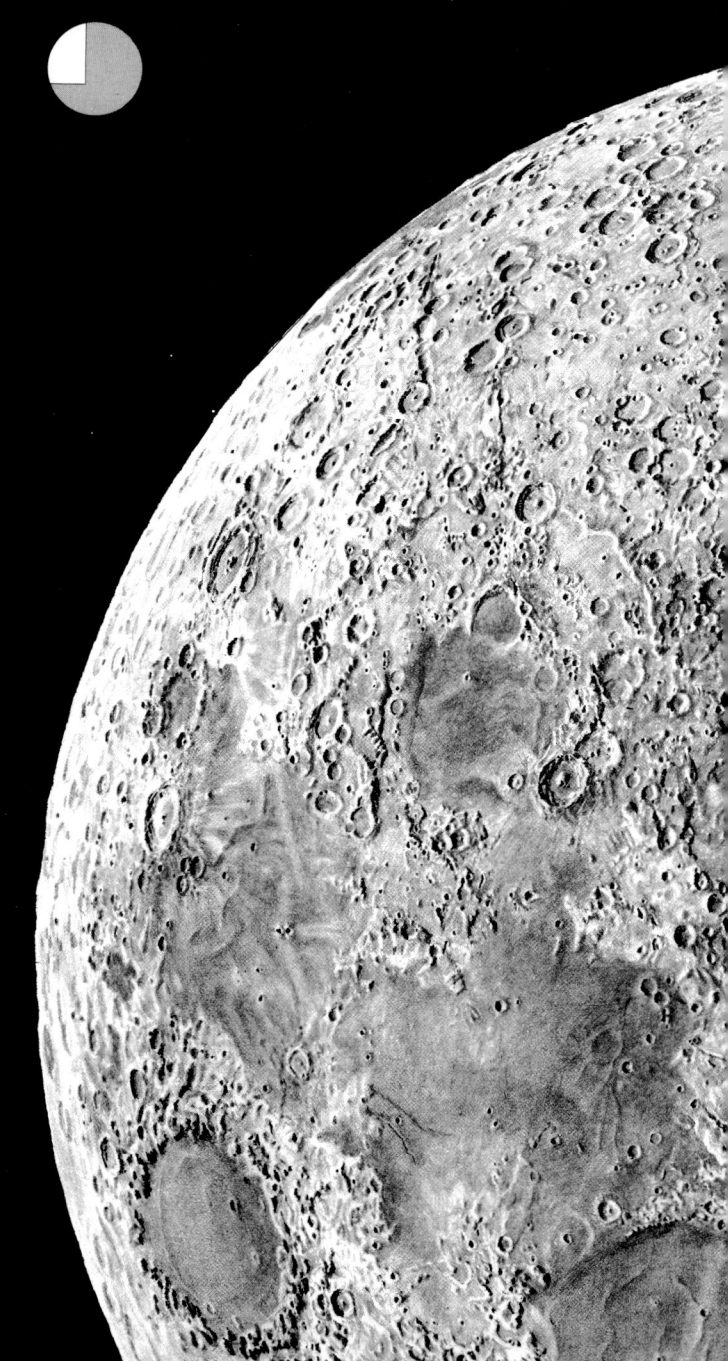

Boussingault
Hagecius
Mutus
Pontécoulant
Nearch
Rosenberger
Hommel
Biela
Vlacq
Watt
Steinheil
Pitiscus
Ideler
Peirescius
Mallet
Lockyer
Barocius
Vega
Fabricius
Oken
Rheita
Janssen
Nicolai
Vallis
Metius
Büsching
Buch
Riccius
Rabbi Levi
Furnerius
Rheita
Stiborius
Lindenau
Zagut
Stevinus
Neander
Rothmann
Adams
Reichenbach
Piccolomini
Snellius
Pontanus
Phillips
Borda
Rupes Altai
Petavius
Wrottesley
Fracastorius
Polybius
Sacrobosco
ataeus
Santbech
Holden
Monge
Beaumont
Catharina
naim
Cook
MARE
Lamé
Vendelinus
Colombo
MONTES
NECTARIS
Tacitus
Magelhaens
PYRENAEUS
Cyrillus
Goclenius
Mädler
Kant
Langrenus
Gutenberg
Gaudibert
Theophilus
Capella
Isidorus
Alfraganus
MARE
Taylor
Torricelli
Delambre
Messier
MARE
FECUNDITATIS
Sabine
Dionysius
MARE
Maskelyne
Ritter
SPUMANS
Secchi
MARE
Apollonius
Taruntius
Arago
Julius
MARE
Rupes
Sinas
Caesar
UNDARUM
Cauchy
TRANQUILLITATIS
Ross
Condorcet
Lyell
Plinius
MARE
Proclus
Vitruvius
Menelaus
MARE MARGINIS
CRISIUM
Dawes
Maraldi
Bessel
Macrobius
Littrow
Plutarch
Römer

MARE AUSTRALE

MARE SMYTHII

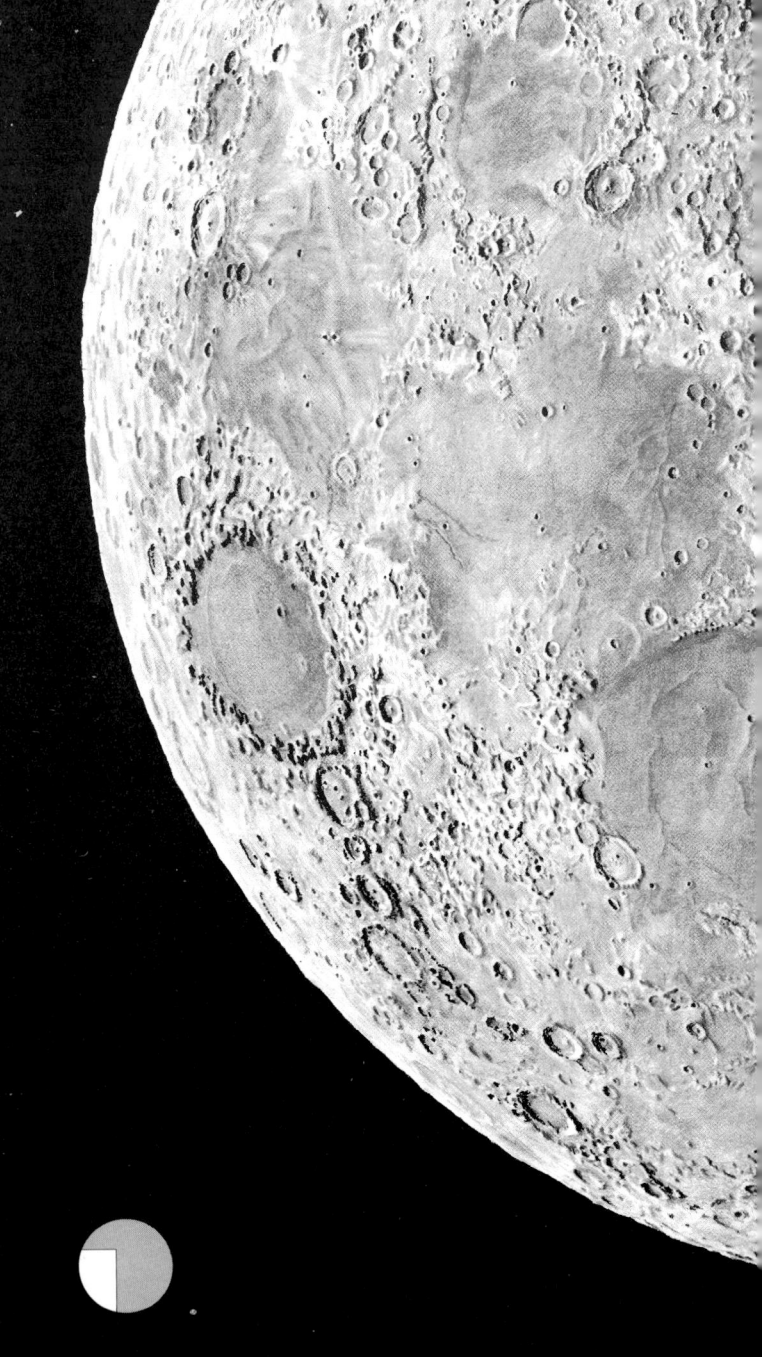

Petavius Wrottesley   Fracastorius   Polybius   Sacrobosco
cataeus   Santbech
                Holden   Monge   Beaumont   Catharina
haim   Cook   Tacitus
        Lamé   Vendelinus   Colombo   MARE   Cyrillus
us   NECTARIS
        Magelhaens   Mädler
use   Goclenius   Theophilus   Kant
                   Langrenus Gutenberg   Gaudibert Isidorus
MARE SMYTHII   Capella   Alfraganus
                MARE   Taylor
                FECUNDITATIS   Torricelli   Delambre
MARE   Messier
SPUMANS   Sabine
                                              Maskelyne   Ritter   Dionysius
        Apollonius   Secchi
MARE   Taruntius   MARE   Arago
UNDARUM   Rupes   Sinas   Julius
                                     Cauchy   Caesar
Condorcet   TRANQUILLITATIS
                                              Ross
MARE MARGINIS
                        Lyell   Plinius
                MARE   Proclus   Vitruvius   Menelaus
                CRISIUM   Dawes
                                     Maraldi   Bessel
                        Macrobius   Littrow
        Plutarch   Römer   MARE
        Cleomedes   SERENITATIS
        Hahn   Tralles   Chacornac
                Burckhardt   Posidonius
        Berosus
                Geminus   Daniell
        Gauss
                        Franklin   LACUS SOMNIORUM
                Messala   Cepheus   Grove
                        Hooke   Plana
                Zeno   Chevallier   Hercules   Bürg
                        Atlas   LACUS
                                        MORTIS
        Mercurius   Aristoteles
                        Endymion
                MARE HUMBOLDTIANUM   Strabo   Democritus

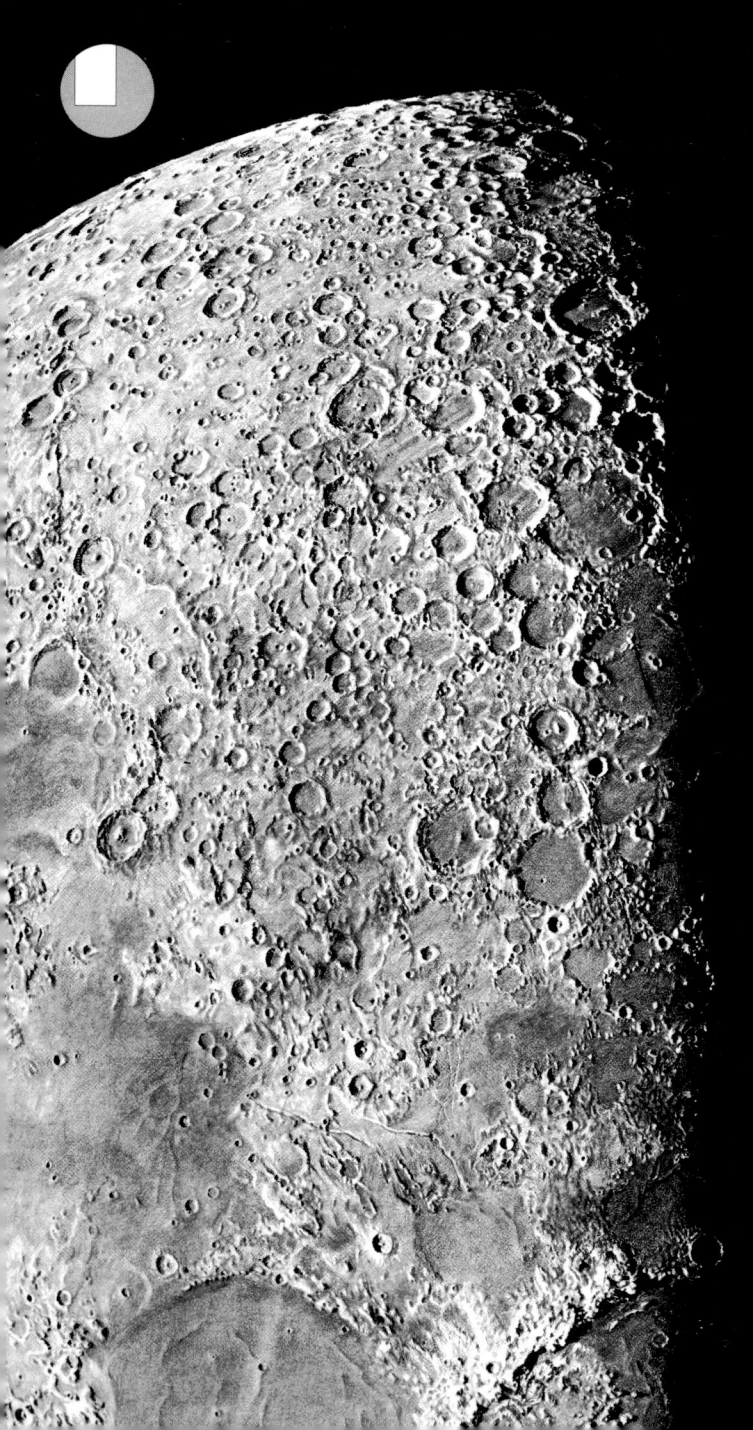

Boguslawsky
Boussingault
Simpelius
Manzinus
Curtius
Hagecius
Mutus
Pentland
Zach
Nearch
Rosenberger
Jacobi
Hommel
Lilius
Vlacq
Baco
Cuvier
Heraclitus
Steinheil
Pitiscus
Ideler
Clairaut
Licetus
Lockyer
Barocius
Faraday
Nasireddin
Janssen
Nicolai
Maurolycus
Stöfler
Miller
Büsching
Buch
Fernelius
Riccius
Kaiser
Rabbi Levi
Gemma Frisius
Lindenau
Walter
Zagut
Goodacre
Rothmann
Aliacensis
Pontanus
Apianus
Werner
Rupes Altai
Sacrobosco
Playfair
Polybius
Azophi
Abenezra
Catharina
Geber
Almanon
Tacitus
Abulfeda
Cyrillus
Descartes
Albategnius
Anděl
Theophilus
Kant
Hind
Halley
Alfraganus
Hipparchus
Taylor
Horrocks
Delambre
Rhaeticus
SINUS
MEDII
Sabine
Maskelyne
Ritter
Dionysius
Godin
Agrippa
Triesnecker
Rima
Arago
Rima Ariadaeus
Hyginus
Ukert
Julius
Caesar
Ross
MARE
VAPORUM
Manilius
Plinius
MONTES
HAEMUS
Menelaus
Dawes
Littrow
Bessel
Conon
MONTES APENNINUS
MARE
PALUS

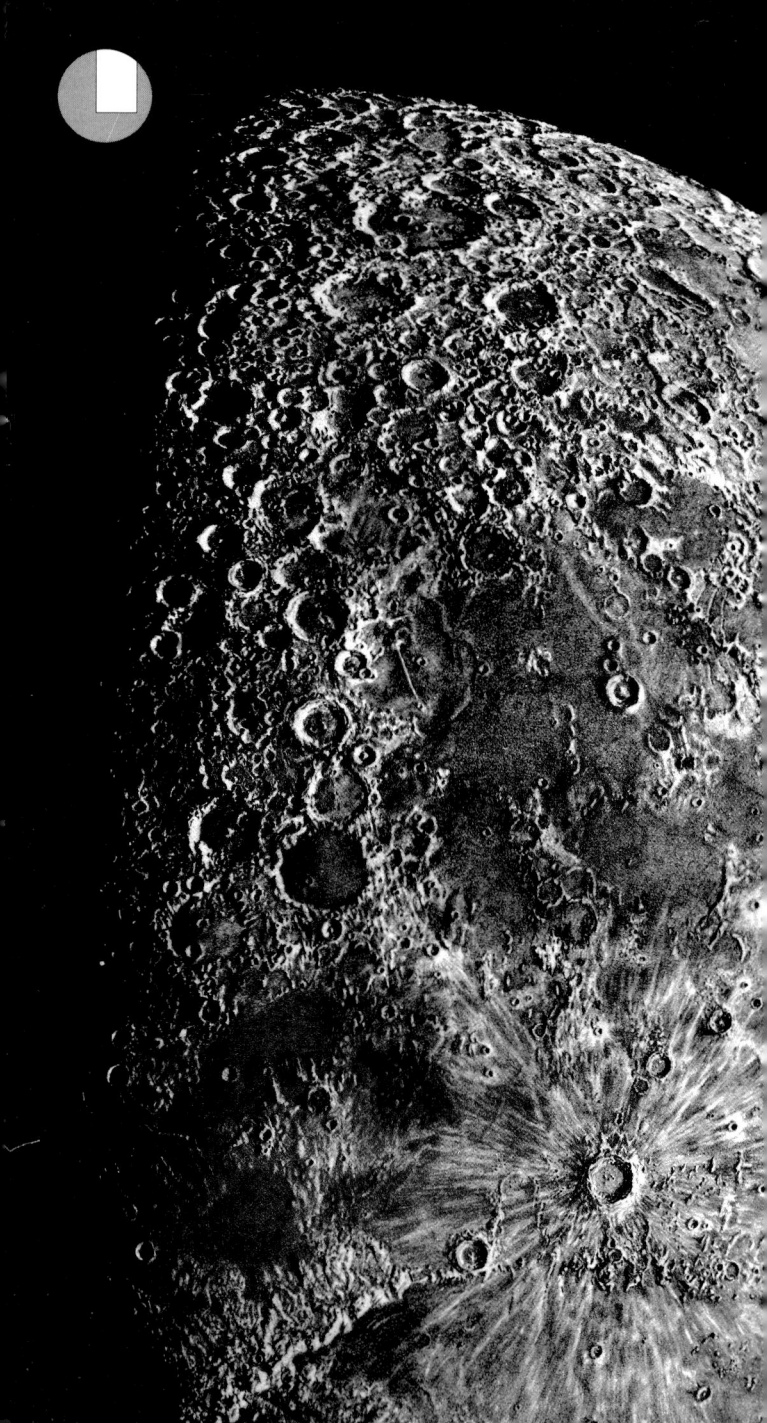

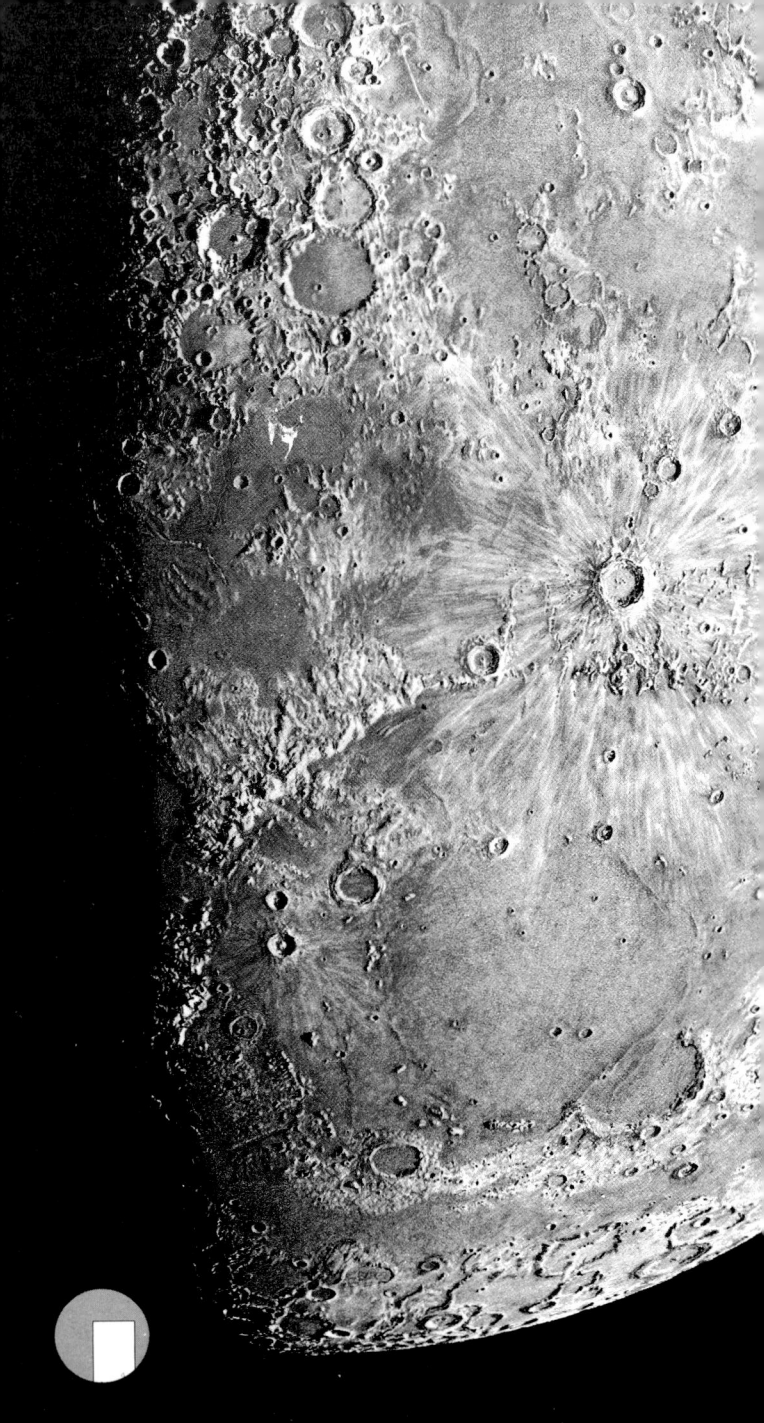

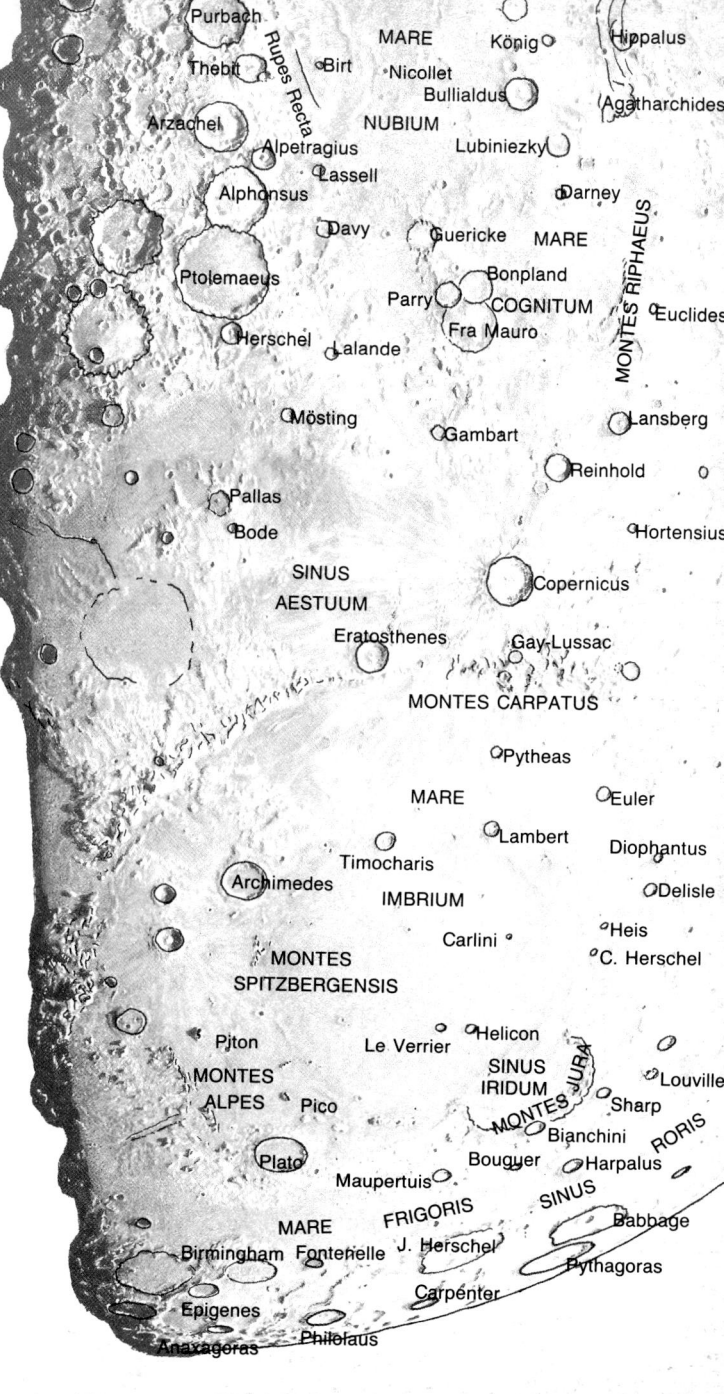

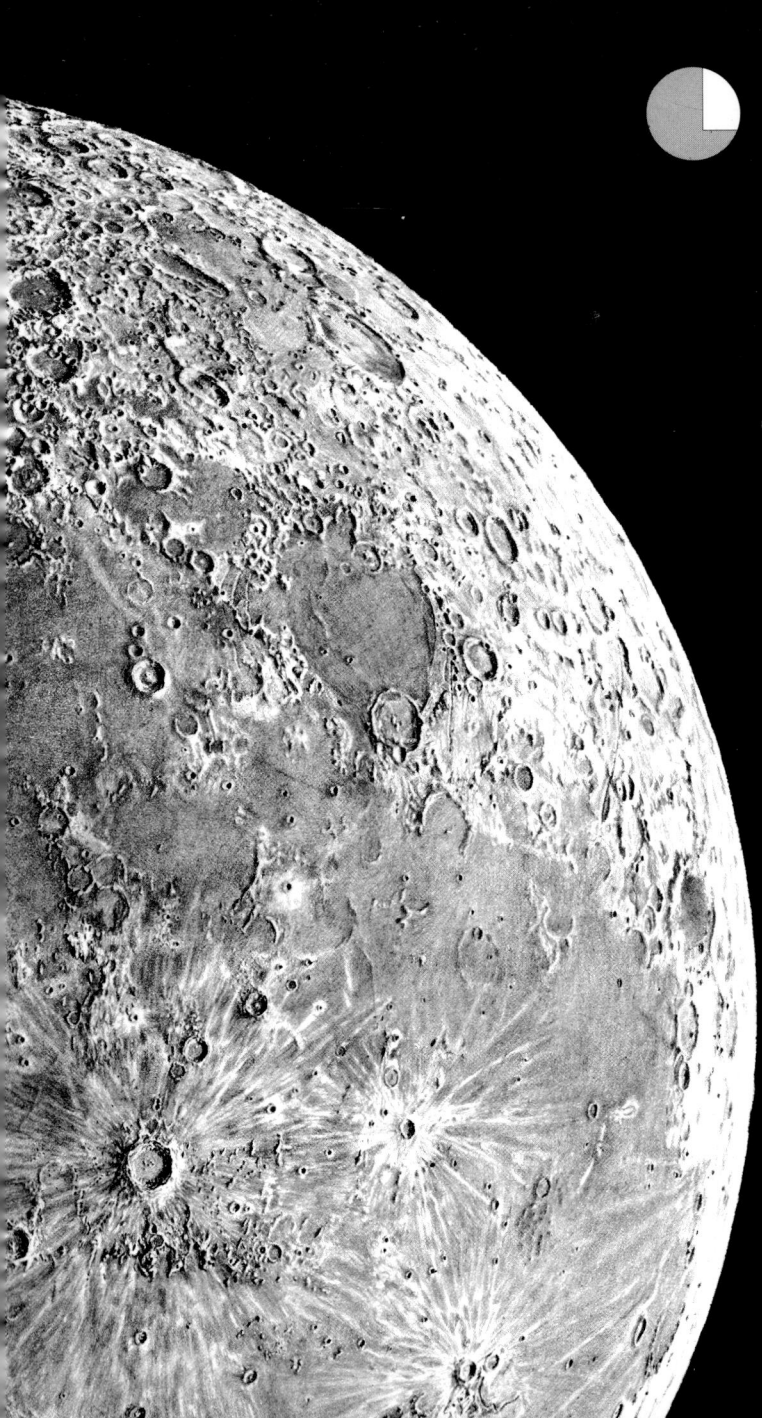

rcher
Bettinus
Zucchius
r
Rost
Phocylides
Schiller
Wargentin
Bayer
Nasmyth
Inghirami
Schickard
m
Mee
Hainzel
Drebbel
auer
Capuanus
ichus
Ramsden
Fourier
Vitello
Vieta
Mercator
Campanus
Doppelmayer
Palmieri
Kies
de Gasparis
Königo
Rimae Hippalus
MARE
Cavendish
aldus
Hippalus
HUMORUM
Rimae Mersenius
Liebig
ubiniezky
Agatharchides
Mersenius
Gassendi
Darney
MONTES RIPHAEUS
Billy
Rima Sirsalis
ericke MARE
Bonpland
Letronne
Hansteen
Sirsalis
COGNITUM
Euclides
ra Mauro
Damoiseau
Grimaldi
Flamsteed
Riccioli
ambart
Lansberg
OCEANUS
Hevelius
Reinhold
Kunowsky
Hedin
Encke
Cavalerius
Hortensius
Kepler
Reiner
Olbers
Copernicus
PROCELLARUM
Marius
Gay-Lussac
Cardanus
Tobias Mayer
Krafft
ES CARPATUS
Seleucus
Pytheas
Struve
Aristarchus
Herodotus
Euler
Schiaparelli
Vallis

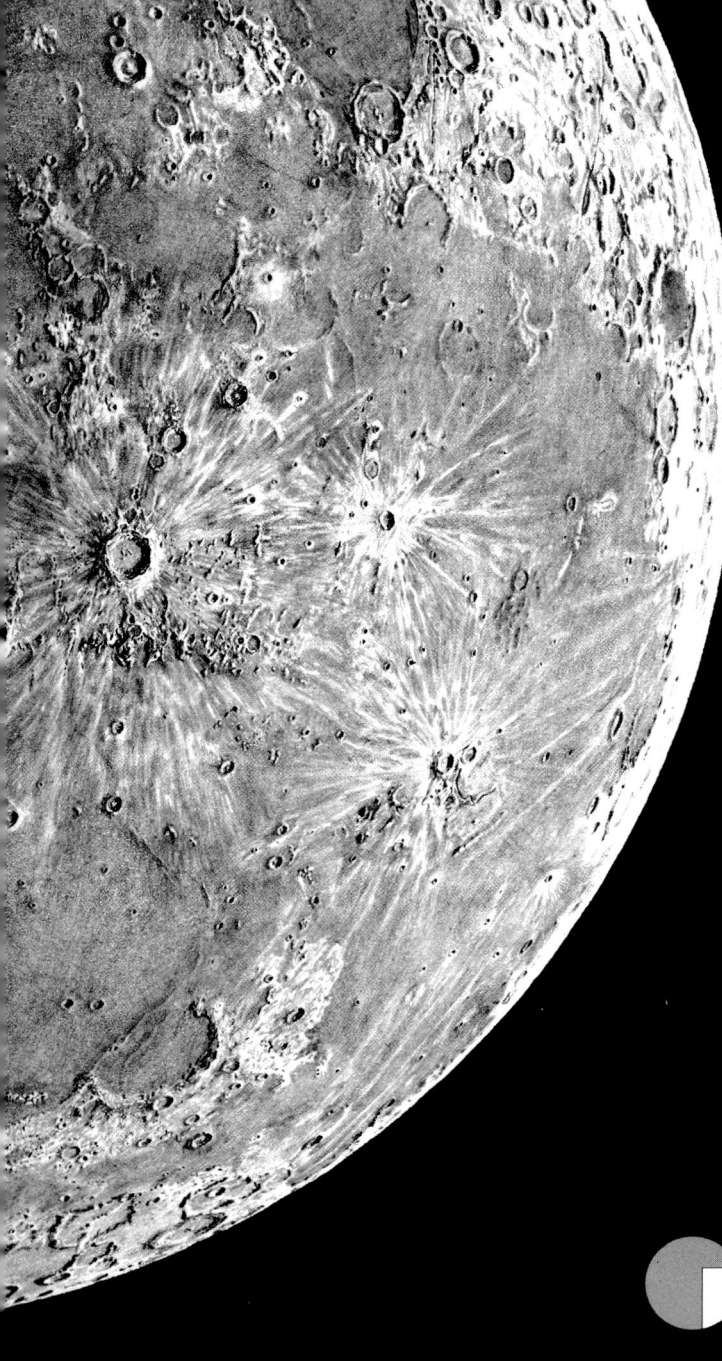

König

HUMORUM

Mersenius

Agatharchides

lialdus

Gassendi

Lubiniezky

Rima Sirsalis

Darney

Billy

Sirsalis

uericke

MARE

Letronne

Hansteen

Bonpland

COGNITUM

Euclides

MONTES RIPHAEUS

Grimaldi

Fra Mauro

Damoiseau

Riccioli

Flamsteed

Gambart

Lansberg

OCEANUS

Hevelius

Hedin

Kunowsky

Reinhold

Cavalerius

Encke

Hortensius

Kepler

Reiner

Olbers

Copernicus

Marius

Cardanus

Gay-Lussac

PROCELLARUM

Krafft

Tobias Mayer

Seleucus

NTES CARPATUS

Pytheas

Aristarchus

Herodotus

Struve

ARE

Euler

Vallis

Schiaparelli

Prinz

Schröteri

Briggs

Lambert

Diophantus

Krieger

RIUM

Delisle

Wollaston

Lichtenberg

Carlini

Heis

C. Herschel

Rümker

Helicon

Mairan

rier

SINUS

Louville

Harding

IRIDUM

MONTES JUR

Sharp

Bianchini

RORIS

Bouguer

Harpalus

Markov

GORIS

SINUS

Babbage

Herschel

Pythagoras

Carpenter

ECLIPSES. From time to time the Earth or the Moon enters the other's shadow, producing an eclipse of either the Sun or the Moon. An eclipse of the Sun occurs at times of new Moon, when the Moon passes between the Sun and the Earth and throws its shadow on the Earth. The Moon itself is eclipsed when it enters the Earth's shadow, which may happen at full Moon. Not every new and full Moon sees an eclipse, because the Moon's orbit is tilted at 5° to the ecliptic. Only when the new or full Moon lies close to the ecliptic do eclipses take place.

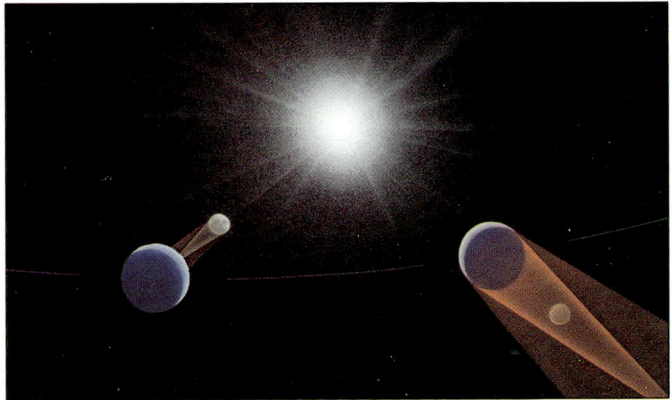

When the Moon's shadow falls on the Earth (left), there is a solar eclipse; lunar eclipses happen when the Moon enters the Earth's shadow (right).

There are at least two eclipses of the Sun each year, and a maximum of five; there can be from zero to three lunar eclipses per year. Not all these eclipses will be visible from any one place on Earth, but most places can expect to experience an eclipse of some kind in most years.

SOLAR ECLIPSES are the most spectacular, and of greatest interest. They are rare for any particular spot on Earth, and astronomers often travel long distances to view them, particularly for the prospect of seeing the pearly *corona*, a faint halo of gases surrounding the Sun. A total solar eclipse is visible only from within the narrow band on which the Moon's *umbra* (the dark central part of its shadow) falls. A partial eclipse is visible from a much wider area within the lighter outer part of the shadow, the *penumbra*.

A solar eclipse begins at *first contact* when the Moon starts to move across the face of the Sun. For the next hour or

A total solar eclipse is the time to see the pearly corona that surrounds the Sun. This is the eclipse of 1980 February 16, photographed from Kenya with an 80 mm lens through 16 × 50 binoculars, exposure 1 second on ISO 64 film.

more a steadily increasing partial eclipse is seen, the progress of which can be followed by projecting the Sun's image, as in normal solar observation. Note the jagged edge of the Moon, caused by crater rims and mountains seen in profile, and compare the jet-black silhouette of the Moon with any sunspots, which will appear brownish in colour.

As totality approaches, the landscape darkens. Ripples of light known as *shadow bands* may be seen on the ground as the last thin crescent of sunlight is refracted through the Earth's atmosphere. Then the crescent is broken up into intense points of light called *Baily's beads* by mountain peaks along the Moon's limb; sometimes one bead is much brighter than the others, producing the *diamond ring* effect.

Totality begins at *second contact* when the Moon completely blots out the Sun's brilliant disk. The Sun's *chromosphere*, a layer of gas above the photosphere, forms a ring of pink light around the Moon. Some prominences may also be visible, as little pink tufts. In the darkened sky the brightest stars and planets are visible. During totality the corona is visible, extending for several solar diameters. Its shape depends on prevailing solar activity, being more regular at solar maximum than at solar minimum.

Totality usually lasts from two to five minutes, the maximum possible being 7½ minutes. It ends at *third contact*, when the Sun starts to reappear, and the Moon finally clears the face of the Sun at *fourth contact*, as much as four hours after first contact.

Total solar eclipses result from the remarkable coincidence that the Sun and Moon appear virtually the same size in the sky. However, when the Moon is at its farthest from the Earth (*apogee*) it is too small to cover the full disk of the Sun. This gives rise to an *annular* eclipse, so named because a ring (or 'annulus') of sunlight remains visible around the Moon. The annular part of an eclipse can last for up to 12½ minutes, but usually is much shorter.

LUNAR ECLIPSES are visible from anywhere provided the Moon is above the horizon at the time. A partial lunar eclipse begins when the edge of the Earth's umbra touches the Moon. This moment is difficult to determine precisely, since the edge of the umbra is diffuse and is preceded by the lighter, outer portion of the Earth's shadow, the penumbra, which itself produces a slight darkening of the Moon. Sometimes the Moon enters only the penumbra, but such *penumbral eclipses* are scarcely noticeable unless the Moon is deep within the penumbra.

The darkness of a total lunar eclipse can be estimated on the following scale, originated by the French astronomer André Danjon, which assigns a luminosity (*L*) value:

L = 0: Very dark eclipse. Moon almost invisible, particularly at mid-totality.
L = 1: Dark eclipse. Grey or brownish coloration, surface details difficult to see.
L = 2: Deep red or rust-coloured eclipse. Very dark centre to the umbra but outer rim relatively bright.
L = 3: Brick red eclipse. Umbra has a bright or yellow rim.
L = 4: Very bright copper-coloured or orange eclipse. Umbral rim is very bright and bluish.

Totality usually lasts well over an hour, but the appearance of the Moon during totality can vary markedly from one eclipse to another. At many eclipses the Moon appears a deep red colour because light from the Sun is being refracted through the Earth's atmosphere. But if the Earth's atmosphere is exceptionally cloudy, or if there is a lot of dust in the upper atmosphere – as there is after a major volcanic eruption, the eclipse can be much darker. Only in exceptional cases does the Moon entirely disappear from view at mid-eclipse.

At the total lunar eclipse of 1989 August 17, the Moon appeared rust-coloured and surface features remained visible. This photograph was taken through a 300mm f/4.5 telephoto lens on a driven mount with a 10 second exposure on ISO 400 film.

THE PLANETS

Planets are a popular subject for observers with small telescopes, and some observations can be made even with binoculars. The brightest planets – Mercury, Venus, Mars, Jupiter and Saturn – can be followed with the naked eye as they orbit the Sun, but Uranus and Neptune require optical aid to be seen. Pluto is so small and faint that it is beyond the reach of small amateur telescopes. Between Mars and Jupiter orbit a host of minor planets, also known as asteroids; the brightest of which can be followed in binoculars and small telescopes. Planets are always found near to the ecliptic.

Whether a planet is visible depends on where it lies in its orbit. Mercury and Venus both orbit between the Sun and the Earth and are known as the *inferior* (or inner) planets. The best time to see Mercury or Venus is near *greatest elongation* when they are at their maximum separation from the Sun (see the illustration). At greatest elongation east they lie in the evening sky. Each planet then moves towards *inferior conjunction*, between the Sun and the Earth, when they cannot be seen because of the Sun's brilliance. Occasionally they pass directly across the face of the Sun, an event known as a *transit*, when they are visible as small black dots. After inferior conjunction Mercury and Venus reappear in the morning sky, passing through greatest elongation west before moving behind the Sun to reach *superior conjunction*, when they are again lost from view.

The *superior* (or outer) planets are best seen around *opposition*, when they are visible all night. At opposition a planet lies opposite the Sun in the sky, so that it rises at sunset and is due south at midnight. As the illustration shows, opposition is also the time when the planet is closest to the Earth and hence appears at its biggest and brightest. For northern hemisphere observers, planets are highest in the sky when opposition occurs in winter, and lowest in the sky

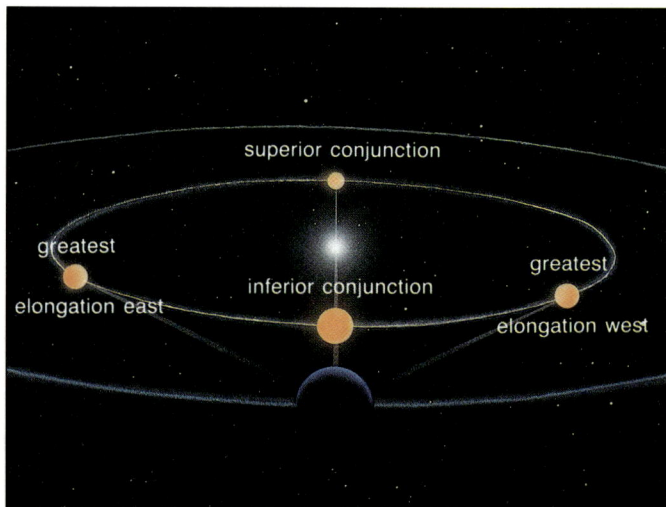

Aspects of an inferior planet.

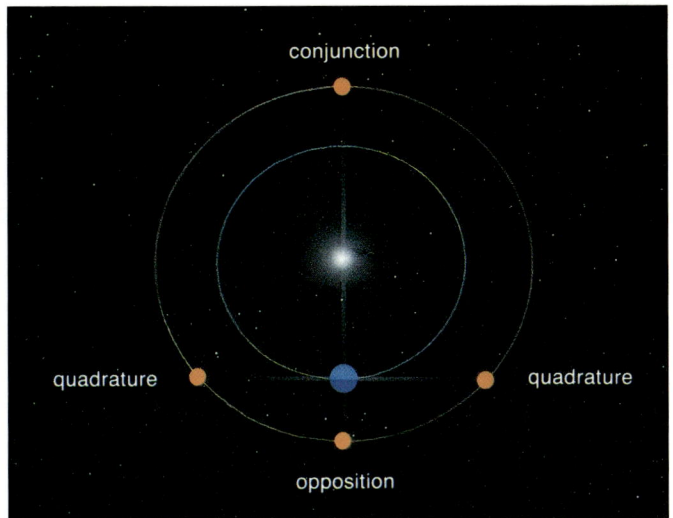

Aspects of a superior planet.

at summer oppositions. The outer planets do not go through phases as do the inner planets, but around the time of *quadrature*, when 90° from the Sun, Mars is noticeably gibbous. *Conjunction* for an outer planet is when it lies on the far side of the Sun from us, in the daytime sky, and is therefore invisible. Outer planets cannot come to inferior conjunction.

All planets orbit the Sun from west to east (anticlockwise as seen from above the Sun's north pole), and their orbital periods increase with increasing distance. A planet's *sidereal period* is the time it takes to go through 360° of its orbit around the Sun relative to the fixed stars. The orbital period relative to the moving Earth, though, is of different duration and is known as the *synodic period*. This, for example, is the time from one elongation to the next, or one opposition or conjunction to the next.

The orbits of the planets are elliptical, and this affects their distances from Earth at successive oppositions. Mars, for example, has particularly close oppositions every 15 years or so. The closest point in a planet's orbit to the Sun is called *perihelion*, and the most distant point is called *aphelion*. A planet's distance from the Sun is often expressed in *astronomical units*, which is the average distance of the Earth from the Sun. One astronomical unit (AU) is 149,6000,000 km.

MERCURY is the closest planet to the Sun and consequently the most elusive, since it is usually drowned in evening or morning twilight. For town-dwellers, simply glimpsing Mercury is a major achievement since a low horizon is necessary. It appears as an orange-coloured star and can reach magnitude 0 or brighter around the time of greatest elongation, but observers in high northerly latitudes will need binoculars to pick it out against the twilight.

MERCURY DATA BOX

Diameter: 4878 km
Mass: 0.06 × Earth
Mean density: 5.43 × water
Volume: 0.06 × Earth
Escape velocity: 4.25 km/sec
Axial inclination: 0°
Sidereal period of axial
 rotation: 58.65 days
Number of moons: 0
Mean distance from Sun:
 57.9 million km
Sidereal period: 88 days
Synodic period: 116 days
Inclination of orbit: 7.0°
Eccentricity of orbit: 0.206

Shadowy markings on Mercury, 1988 November 3, 150 mm refractor, ×286.

There are six (or sometimes seven) elongations of Mercury each year, but they are not all equally favourable. From the northern hemisphere, Mercury is highest in the sky at evening elongations in the spring and at morning elongations in the autumn. The planet can be studied for about a week either side of greatest elongation.

Small telescopes show that Mercury goes through a cycle of phases as it orbits the Sun. Little else is visible since the planet is so small, only 40% larger than the Moon, and it is usually viewed close to the horizon where the seeing is bad. At greatest elongation a magnification of over 200 times is needed to make Mercury appear as large as the full Moon does to the naked eye. In 1974 the US Mariner 10 space probe sent back photographs which showed that Mercury looks very much like our own Moon. Under favourable conditions some dusky surface markings may be visible on Mercury with apertures as small as 100 mm (4 inches), but for serious study large apertures are required. In addition to drawing any surface markings, observers should make estimates of the phase of Mercury and look for irregularities in the terminator.

Occasionally Mercury can be seen crossing the face of the Sun. These transits happen when Mercury reaches inferior conjunction at the same time as it crosses the plane of the Earth's orbit, which it does in November and May. On such occasions the planet is visible as a tiny black dot silhouetted against the face of the Sun for several hours. Timings are required of the entry of the disk of Mercury onto the Sun's disk, which takes about two minutes, and its subsequent exit. At entry, look for the so-called *black drop*, an apparent lingering connection between the disk of Mercury and the limb of the Sun. Forthcoming transits of Mercury will be on 1993 November 6 (mid-transit 4 hours UT), 1999 November 15 (22h), 2003 May 7 (8h) and 2006 November 8 (22h).

VENUS is probably the easiest planet to identify, because of its extreme brilliance, but since it is so brilliant it is best observed against a bright sky to cut down the amount of dazzle. Venus is almost the same size as the Earth and its orbit is closer to ours than that of any other planet. Its brilliance is due both to its closeness and to its high reflectivity, or *albedo*, for the planet is entirely surrounded by clouds that reflect two-thirds of the incoming sunlight.

Y-shaped dusky cloud markings and irregular terminator of Venus, 1972 March 22, 215mm reflector, ×220.

As with Mercury, Venus appears in the evening or morning sky but it can be seen over a much wider arc of its orbit than Mercury. Its elongations are much greater than those of Mercury, up to 47°, and at best it can be observed for several hours after sunset or before sunrise. At greatest elongation a magnification of 75 times will show Venus the same size as the Moon appears to the naked eye. Venus reaches its greatest brilliancy, up to magnitude −4.7, halfway between greatest elongation and inferior conjunction. When Venus is a crescent it appears so large that even binoculars will show its phase.

Because of its unbroken cloud cover, the surface of Venus can never be seen through telescopes from Earth. However, some cloud features can be seen from time to time, most noticeably brighter patches around the poles of the planet,

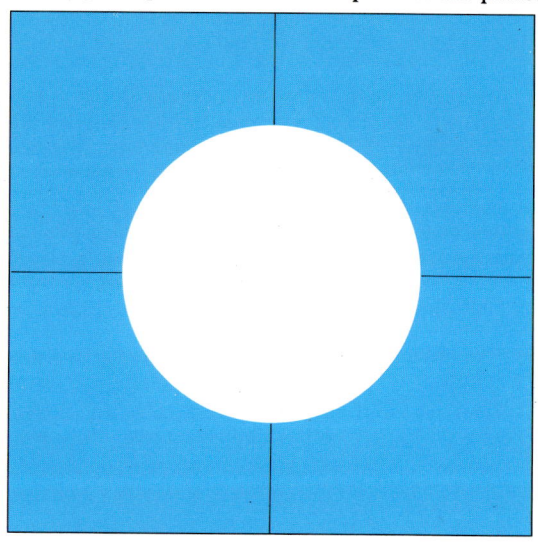

Venus observing report form

Observer and location: ....
Date and time (UT): ....
Aperture and magnification: ....
Seeing and transparency: ....
Predicted and observed phase (%): ....
Notes and intensity estimates (0 to 5): ....
....
....
....

	Definitely present	Possibly present	Definitely absent
N. cusp cap	☐	☐	☐
N. cusp cap collar	☐	☐	☐
S. cusp cap	☐	☐	☐
S. cusp cap collar	☐	☐	☐
Terminator shading	☐	☐	☐
Terminator irregularity	☐	☐	☐
Limb brightening	☐	☐	☐
Cusp extensions	☐	☐	☐
Ashen light	☐	☐	☐
Night side darker	☐	☐	☐

termed *cusp caps* since they lie at the cusps, or horns, of the crescent planet. Often there is a dark band, or *collar*, around the cusp caps. The reality of these features is confirmed by space probe photographs of the planet's clouds.

The clouds rotate around Venus every four days, but the planet itself rotates on its axis very much more slowly, every 243 days. Both motions are *retrograde*, i.e. from east to west, opposite to the direction of spin of the Earth and other planets (except the highly tilted Uranus). Beneath the clouds Venus is a roastingly hot, rocky world with two major upland areas and signs of volcanic activity.

Observers should draw the planet's disk to a standard diameter of 50 mm; a sample blank is included on the report form shown here. The simplest observation to make of Venus is its phase, which can be estimated by measuring your drawing with a ruler. Surprisingly, around *dichotomy* (half-phase) the observed phase does not tally with the predicted phase. Eastern (evening) dichotomy is usually a few days early, and western (morning) dichotomy a few days late. This is known as *Schröter's effect*, and may be a result of the terminator being less bright than the rest of the planet.

The planet's terminator can appear somewhat irregular, including extensions or blunting of the cusps, apparently due to cloud effects. Elusive dusky cloud shadings can be seen on the planet, often shaped like a sideways V or Y. Notes should be made of all features recorded, in the space provided on the report form. On your drawing you will need to exaggerate the darkness of cloud features for clarity. The intensity of features on Venus is estimated on the following scale:

   0: extremely bright areas such as white spots
   1: bright areas such as cusp caps
   2: general hue of the disk
   3: elusive shading near the limit of visibility
   4: shading well seen
   5: unusually dark shading

VENUS DATA BOX
Diameter: 12,100 km
Mass: 0.82 × Earth
Mean density: 5.24 × water
Volume: 0.86 × Earth
Escape velocity: 10.36 km/sec
Axial inclination: 177°
Sidereal period of axial rotation: 243 days (retrograde)
Number of moons: 0
Mean distance from Sun: 108.2 million km
Sidereal period: 224.7 days
Synodic period: 583.9 days
Inclination of orbit: 3.4°
Eccentricity of orbit: 0.007

A curious effect often mentioned is the *ashen light*, an apparent brightening of the night side of the crescent Venus when seen against a dark sky, which may be caused by aurorae in the planet's atmosphere. However, to observe the ashen light reliably the illuminated portion of the planet must be blocked out by covering part of the eyepiece with a strip of paper or metal; without this precaution any reports will probably be spurious. Conversely, in daylight or twilight the night side of the planet can seem darker than the surrounding sky, for reasons that are not fully understood.

Transits of Venus across the face of the Sun are even rarer than those of Mercury. The next two will take place on 2004 June 7 and 2012 June 5; they will be the first pair of transits for over a century and the last for another century. Timings of the passage of Venus onto the disk of the Sun, and its subsequent exit, will be required, as well as reports of the *black drop* effect, which is a dark 'neck' between the disk of the planet and the limb of the Sun at entry and exit. Look for signs of sunlight refracted around the planet by its atmosphere.

MARS is potentially the most exciting planet of all to observe. It is a rocky world, about half the size of Earth, and undergoes seasonal changes on its surface including melting of its polar caps and variations in the size and shape of dark areas. Its atmosphere is so thin that we can usually see straight through to the surface, but occasionally clouds and even enormous dust storms interrupt our view. Mars studies are all the more important because one day humans will walk on its surface.

Mars comes to opposition every 26 months, but some oppositions are much more favourable than others because of the significant ellipticity of its orbit (see the illustration). When oppositions occur at perihelion, as in 1988 and 2003, Mars can come as close as 56 million km, when it shines as a brilliant orange star of magnitude −2.8. A magnification of 75 times will then show it the same size as the full Moon is to the naked

---

MARS DATA BOX

Diameter: 6787 km
Mass: 0.11 × Earth
Mean density: 3.94 × water
Volume: 0.15 × Earth
Escape velocity: 5.02 km/sec
Axial inclination: 25.2°
Sidereal period of axial
  rotation: 24h 37m
Number of moons: 2
Mean distance from Sun:
  227.9 million km
Sidereal period: 687 days
Synodic period: 780 days
Inclination of orbit: 1.9°
Eccentricity of orbit: 0.093

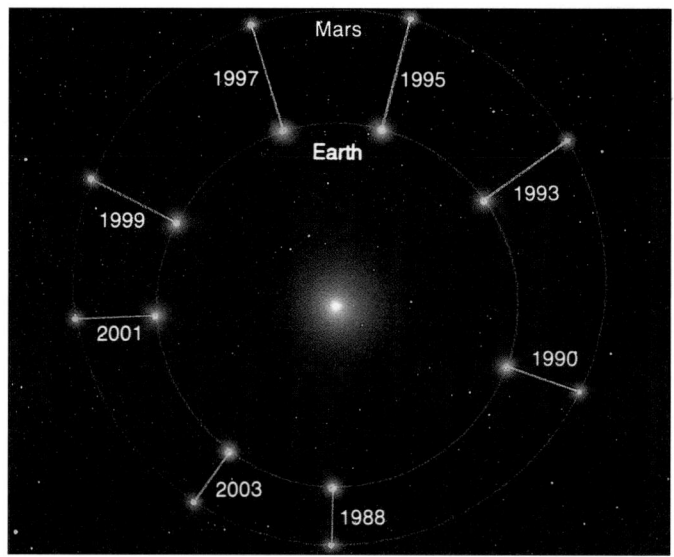

Oppositions of Mars 1988 – 2003.

eye. But at an aphelic opposition, as in 1995, Mars lies 100
million km away and needs a magnification of 130 times to
bring it up to the size of the naked-eye full Moon.

Through a telescope Mars shows an orange disk with white
polar caps and some dark surface markings, the orange colour
being produced by iron oxide in the surface rocks. The polar
caps are composed of a mixture of frozen water and frozen
carbon dioxide. They shrink noticeably during summer in
each hemisphere, leaving detached 'islands'. The south polar
cap shrinks more than the north polar cap because the
southern hemisphere of Mars is tilted towards the Sun at
perihelic oppositions and so gets warmer than the northern
hemisphere, which is tilted towards the Sun at aphelic
oppositions.

The dark markings on Mars were once thought to be due
to vegetation, but are now known to be areas of rock and
dust. These markings change in appearance during the year as
winds blow the dust around, so any map of Mars can be only
an approximation to the planet's true appearance. Perhaps the
most famous marking on Mars is a dark wedge-shaped feature
on the equator called Syrtis Major. Space probes have shown
this to be a sloping area of bare rock. South of it is a bright
circular area called Hellas where white and yellow clouds

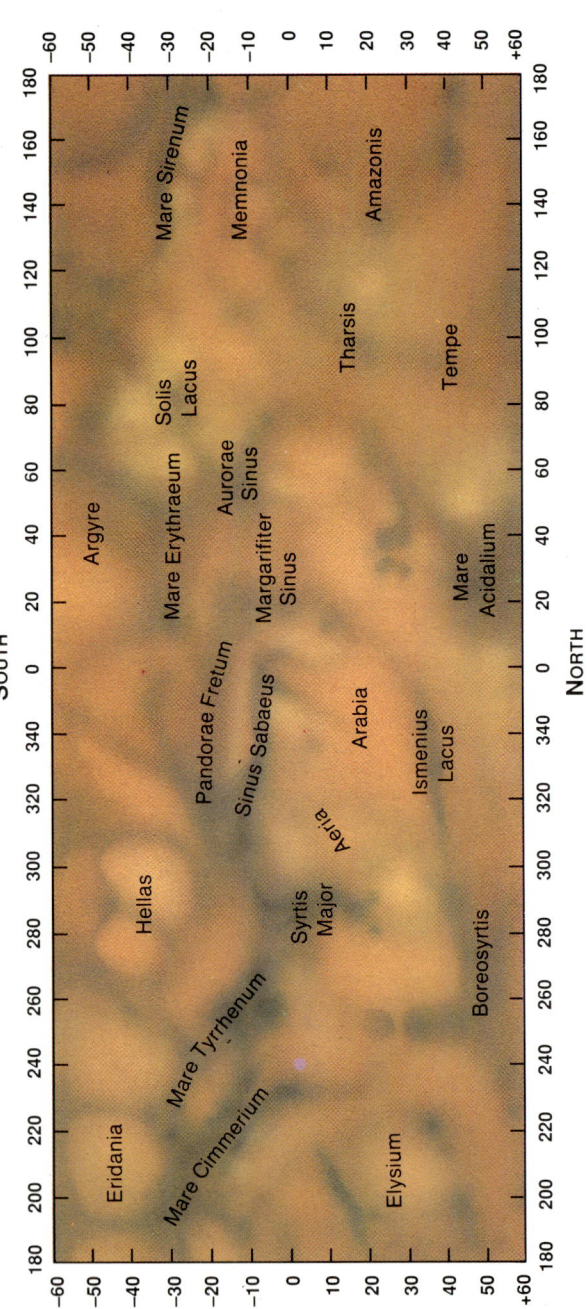

Map of Mars as seen by observers on Earth. South is at the top.

form. Hellas is in fact a large impact basin, similar to the maria on the Moon, but filled with light-coloured dust.

On the opposite side of the planet from Hellas is a bright circular feature called Nix Olympica ('the snows of Olympus'), now also known as Olympus Mons (Mount Olympus) as space probes have shown it to be a huge volcano, the largest of a family of volcanoes in this area. White clouds of water-ice crystals often form around these volcanic mountains. Another major feature discovered by space probes is a huge rift valley 4000 km long. This is filled with dark dust and is visible from Earth as an elongated feature called Coprates. Observers in the past reported seeing fine lines crossing the planet, which they called 'canals'. These canals of Mars are illusory, apparently a result of the eye connecting disjointed markings.

Sometimes yellow clouds are visible on the planet, particularly in the areas of Hellas and Solis Lacus. These clouds are in fact dust storms which, with the increased temperatures at perihelion, can be whipped up by high winds to envelop the entire planet and obscure its features, as happened in 1971 and 1973.

To draw Mars, a standard disk size of 50 mm is used by European observers (you can use the same blank as for Venus on page 78) but American amateurs prefer a 42 mm disk, which corresponds to the planet's diameter of 4200 miles. Whichever size you choose, fill in any phase (Mars can appear only 84% illuminated at quadrature) and then draw the polar caps. Position the major dark markings and finish by adding finer details, including any bright clouds. Advanced observers use filters of different colours to make the various features

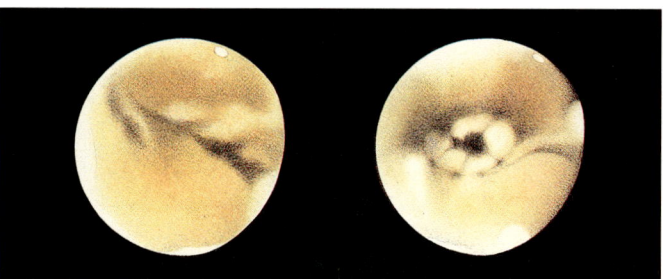

Two views of Mars during its close approach in 1988, 115mm refractor, ×186. Left, November 4, c.m. 189°, showing the elongated Mare Cimmerium; and, right, November 14, c.m. 98°, showing the eye-like Solis Lacus.

stand out more prominently. The intensity of features should be estimated on a scale of 0 for the polar caps to 10 for the blackness of the night sky; the brightest deserts will rate about 2 on this scale and the darkest surface markings about 8. You should also note the longitude of the central meridian, c.m. (i.e. the longitude on Mars that lies at the centre of the disk), which can be found from an annual handbook. Mars turns through nearly 15° in an hour.

Mars has two tiny moons, Phobos and Deimos, both of which are believed to be captured asteroids, but they are too faint and too close to the planet to be seen in small apertures.

JUPITER is the largest of the planets, and is one of the most popular subjects for amateur observation. Every 13 months it reaches opposition, when it can shine as brightly as magnitude −2.9, and a magnification of only 40 times will show it as large as the full Moon appears to the naked eye. But even well away from opposition, humble binoculars will reveal its disk and its four brightest moons, which look like faint stars strung out in a line either side of the planet.

Jupiter is not a solid, rocky world like the Earth − it is basically a huge ball of liquid hydrogen, wrapped around by alternating bright zones and dark belts of cloud that are named according to their latitude (equatorial, tropical, temperate or polar). The dark belts lie at a lower level in Jupiter's atmosphere than the bright zones. All the cloud features on Jupiter are continually changing in appearance, providing a perennial source of interest for amateur observers.

Even a glimpse of Jupiter through a small telescope will reveal two equatorial belts, one to the south and one to the north of the bright equatorial zone. Jupiter's disk is noticeably flattened by its speed of rotation, which is faster than any other planet in the Solar System. The region between the middle of the North and South Equatorial Belts, known as System I, rotates on average every 9h 50m 30s. The

---

JUPITER DATA BOX

Diameter: 42,800 km
Mass: 318 × Earth
Mean density: 1.33 × water
Volume: 1323 × Earth
Escape velocity: 59.6 km/sec
Axial inclination: 3.1°
Sidereal period of axial rotation:
  9h 50m 30s (System I)
  9h 55m 41s (System II)
Number of moons: 16
Mean distance from Sun:
  778.3 million km
Sidereal period: 11.86 years
Synodic period: 399 days
Inclination of orbit: 1.3°
Eccentricity of orbit: 0.048

---

average rotation period of the rest of the planet, termed System II, is five minutes longer. Hence the equatorial regions gain on the regions to the north and south by one rotation every 48 days.

Careful inspection of the various belts reveals numerous loops and plumes of cloud along their edges which change in appearance almost daily, while the belts and zones themselves change in colour and intensity over periods of months or years. The disk of Jupiter is usually a creamy colour, but various cloud features can appear brown, pink or even blue, depending on their composition.

The nearest thing to a permanent feature on Jupiter is the Great Red Spot, an eye-shaped cloud that varies between 24,000 km and 50,000 km in length in the South Tropical Zone, indenting the South Equatorial Belt. Despite its name the spot is not always red in colour: often it is a subtle pink and at times it can fade away completely, leaving a colourless oval known as the Red Spot Hollow, as happened in the mid-1980s. Yet despite all these changes it has persisted since the first telescopic observations of the planet were made over 300 years ago. Of course, the rotation of Jupiter will periodically carry the Red Spot out of view, so observers should not expect to see it every time they look at the planet.

Other spots, both light and dark, erupt on Jupiter from time to time, and the most useful type of observation that amateurs can make is to time their rotation period. This is done by noting when the object crosses Jupiter's central

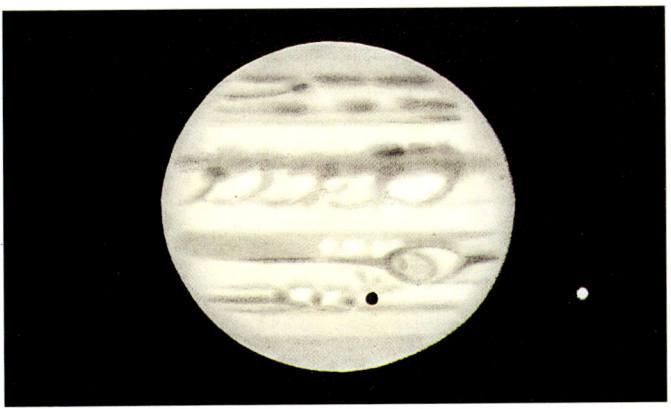

Jupiter on 1987 November 27, 150mm refractor, ×222. The moon Europa casts its shadow onto the planet's clouds, near the Great Red Spot.

meridian, an imaginary line joining the north and south poles of the planet. Separate timings can be made for the leading edge, centre and trailing edge of a spot, and these timings can be converted into Jovian longitudes by reference to tables. Accuracies of a degree in longitude can be achieved by timing to the nearest minute. Successive timings will reveal changes in the spot's longitude caused by winds in Jupiter's turbulent atmosphere.

Jupiter should be drawn with an equatorial diameter of 64 mm and polar diameter 60 mm, as on the observing report form reproduced here. Sketch in the main cloud belts

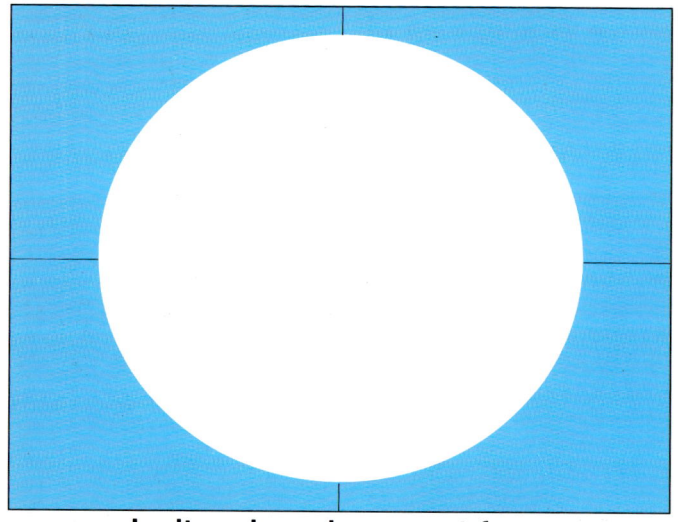

Jupiter observing report form

Observer and location: . . . .

Date and time (UT): . . . .

Aperture and magnification: . . . .

Seeing and transparency: . . . .

Longitude of c.m. System I: . . . .

Longitude of c.m. System II: . . . .

Notes and intensity estimates: . . . .

. . . .

. . . .

. . . .

. . . .

. . . .

quickly, then add detail starting at the leading limb where the rotation of the planet will carry the features out of sight rapidly – Jupiter rotates through 6° every 10 minutes. Being gaseous, Jupiter shows noticeable limb darkening so features close to the limb will not be as well defined as those at the centre. Intensities of various features can be estimated on a scale from 0 (the brightest) to 10 (black sky), as used in Europe, or the opposite (0 dark, 10 brightest) in the USA.

Jupiter's four largest moons – Io, Europa, Ganymede and Callisto – provide almost as much entertainment as the planet

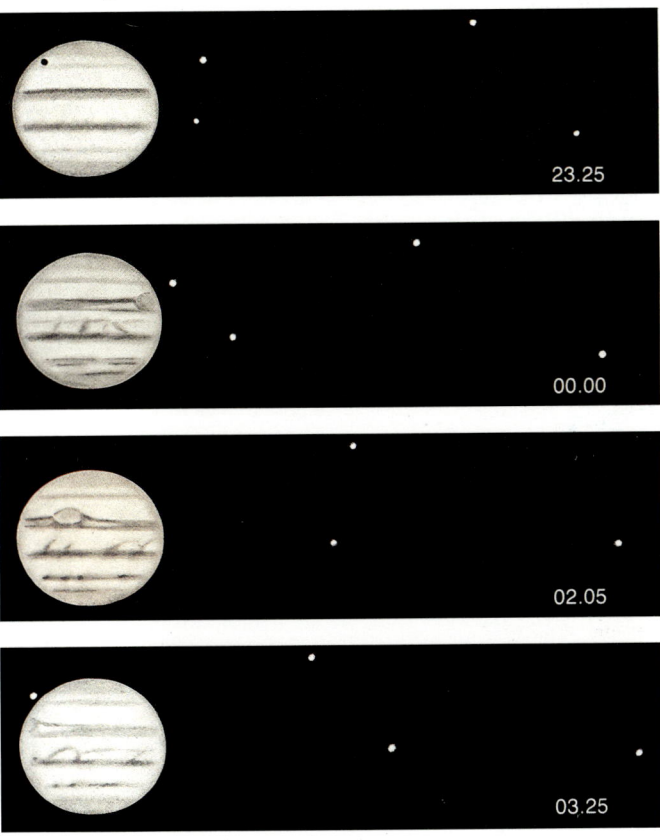

The dance of Jupiter's moons over four hours on the night of 1987 September 5/6. Ganymede moves across the face of the planet from right to left. Below it Io moves away, passing Callisto moving towards Jupiter. Europa is nearly stationary on the right.

itself, particularly for users of small instruments. They appear like fifth- and sixth-magnitude stars, changing their positions nightly as they orbit the planet in periods ranging from less than two days (Io, the closest) to over a fortnight (Callisto, the most distant). The four are known as the Galilean moons, because they were discovered by Galileo.

Sometimes one or more of them may be missing from view, either eclipsed in Jupiter's shadow or hidden behind its disk. Regularly the Galilean moons can be seen in transit across the face of Jupiter, throwing their own small, rounded shadows onto the planet's clouds. Before opposition, a transiting moon is preceded by its shadow; after opposition, the shadow follows the moon. Timings of transits, eclipses and occultations of the satellites can help refine our knowledge of the orbits of these small worlds.

SATURN. Seen through a telescope, Saturn and its encircling rings are widely regarded as the most stunning sight in the sky. Space probes have now found rings around Jupiter, Uranus and Neptune, but the rings of Saturn are the only ones bright enough to be visible from Earth.

SATURN DATA BOX

Diameter: 120,000 km
Mass: 95.2 × Earth
Mean density: 0.7 × water
Volume: 752 × Earth
Escape velocity: 35.6 km/sec
Axial inclination: 26.7°
Sidereal period of axial
  rotation: 10h 14m (equatorial)
Number of moons: 20 +
Mean distance from Sun:
  1427 million km
Sidereal period: 29.46 years
Synodic period: 378 days
Inclination of orbit: 2.5°
Eccentricity of orbit: 0.056

At opposition Saturn appears like a bright yellowish star; its exact magnitude depends on the angle at which the rings are tilted towards us (see opposite). A magnification of 90 times will show the planet itself as large as the full Moon viewed with the naked eye. The rings, though, are far larger than the planet, spanning 275,000 km from rim to rim.

Saturn is similar in nature to Jupiter, being a ball of liquid hydrogen surrounded by clouds, but its cloud features are far less prominent because they are masked by haze in the planet's upper atmosphere. Through a telescope Saturn appears as an ochre-coloured ball, crossed by darker belts and lighter zones that follow a similar pattern to those on Jupiter. There is no equivalent of the Great Red Spot, although various white spots are visible from time to time. Saturn throws its shadow onto the rings, and the rings throw their shadow onto the planet creating the impression of a dark equatorial belt.

As with Jupiter, the main contribution that amateurs can make to serious studies of Saturn is to time the transit of any visible cloud features across the central meridian, identifying each feature on a drawing of the planet. Saturn's equatorial region has an average rotation period of around 10h 14m, about 24 minutes faster than at higher latitudes, but there is not such a clear-cut distinction in the rotation periods as there is with System I and System II on Jupiter.

Even though Saturn spins more slowly than Jupiter, it is more flattened in shape because it has a much lower density. Its density, less than that of water, is the lowest of any planet in the Solar System. Saturn's elliptical outline is not immediately noticeable since the eye is distracted by the rings that encircle the planet's equator.

The equator of Saturn is tilted at nearly 27° to the plane of its orbit, and as the planet moves around the Sun we see the rings presented to us at an angle that varies from 0° to over 27° (the maximum possible angle is somewhat greater than 27° since the slight tilt of Saturn's orbit relative to ours must also be taken into account). When the rings are fairly wide open, their elliptical outline can be seen in binoculars. But when the rings are presented edge-on to us they disappear from view in even the largest telescopes, as will happen in 1995–6.

Saturn with rings wide open on 1989 July 3, 150mm refractor, ×286.

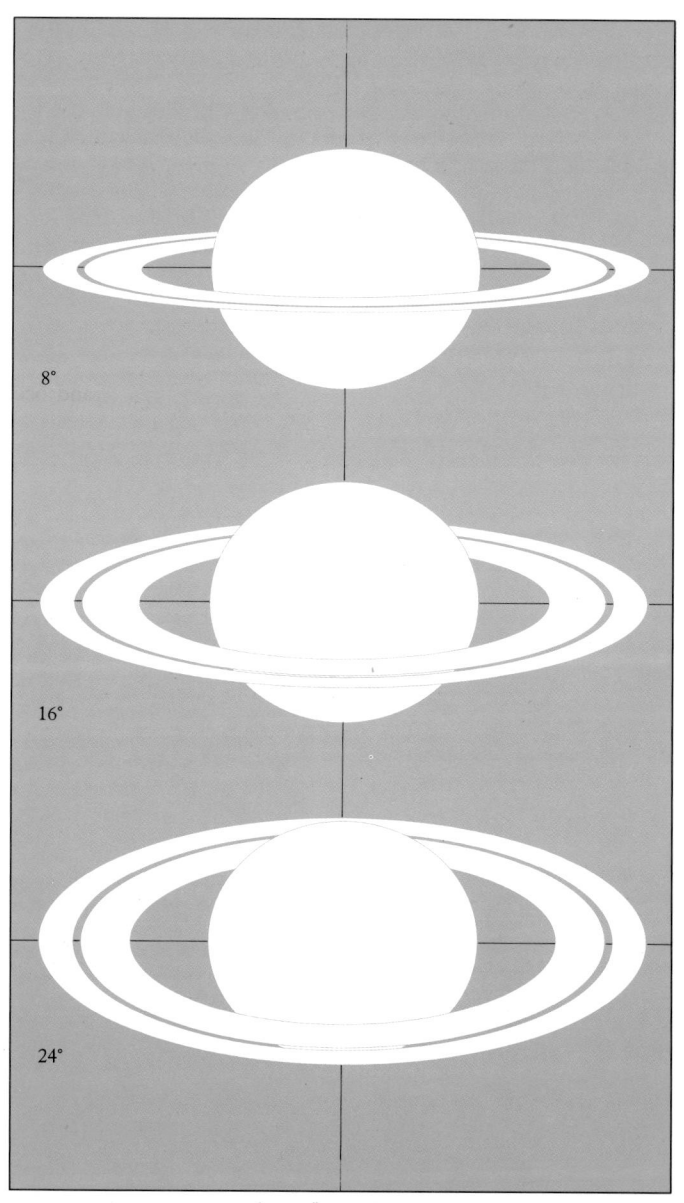

8°

16°

24°

Blanks for drawing Saturn, with the rings presented at various angles.

Countless tiny, ice-covered moonlets make up the rings of Saturn. A telescope of about 75 mm (3 inch) aperture will reveal a dark gap in the rings, the Cassini Division, 3500 km wide. Outside the Cassini Division is the part of the rings known as Ring A, which itself is split by a narrower gap, the Encke Division, visible in larger amateur telescopes. Inside the Cassini Division lies Ring B, the brightest part of the rings. Between Ring B and the planet lies the faintest part of the rings, known as Ring C or the Crêpe Ring because of its transparency. It is too faint to be seen directly in the smallest telescopes, but it may appear as a dusky band where it passes in front of the planet.

Saturn should be drawn in a similar way to Jupiter, quickly marking in the main features and then adding details before the planet's rotation takes them out of view. Because of the changing aspect of the rings a single outline blank for Saturn is inadequate; shown here is a series of three for different inclinations. A useful line of research is to make intensity estimates of the zones and belts of the planet and the rings. Two scales are in use: European astronomers use a scale from 1 (Ring B) to 10 (shadow or black sky) whereas the US scale runs the other way, from 8 (Ring B) to 0 (black sky).

Saturn's largest moon, Titan, is of eighth magnitude and is visible in small telescopes as it orbits Saturn every 16 days. It is of particular interest because it is larger than the planet Mercury, and is the only moon in the Solar System to have a dense, cloudy atmosphere.

URANUS, NEPTUNE AND PLUTO

URANUS, at magnitude 5.5, is in theory visible to a sharp eye on a clear night, but it was unknown until discovered telescopically in 1781 and is not regarded as one of the naked-eye planets. Binoculars will show it clearly as a starlike point of bluish cast, but even at its closest a magnification of over 450 times is needed to enlarge it to the apparent size of the full Moon.

Uranus is so small and remote that amateurs can achieve little in the way of serious studies. Even through large telescopes the planet appears merely as a blue–green, featureless disk, an impression that was confirmed by the Voyager 2 space probe that flew past it in 1986. With the aid of a suitable chart its movement from night to night can be followed in binoculars, and its brightness estimated by comparison with the background stars. Amateur observers

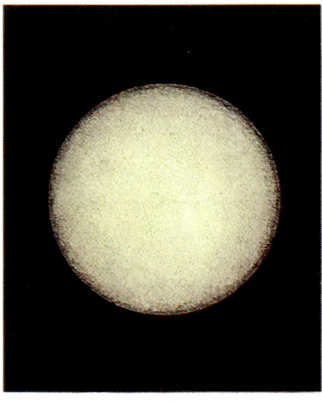

The light green disk of Uranus, 1988 June 13, 150mm refractor, ×333.

have reported changes in the brightness of Uranus, apparently due to the effects of solar activity on its atmosphere.

Uranus is unusual in that its axis of rotation lies almost in the plane of its orbit, so that in the course of its 84 year revolution period both poles are presented in turn towards the Sun. None of its moons is bright enough to be visible in small telescopes.

The aquamarine disk of Neptune, 1989 July 4, 150mm refractor, ×333.

NEPTUNE, at eighth magnitude, is visible in good binoculars, but it is not an easy object. Like Uranus it can be followed as it moves against the stars from night to night, and its magnitude can be estimated for signs of variability caused by solar activity.

Through a telescope it is even more disappointing than Uranus, being half the apparent size and appearing as only a bluish speck even under high power. Its largest moon, Triton, is of 13th magnitude – too faint for most amateur telescopes.

NEPTUNE DATA BOX

Diameter: 48,600 km
Mass: 17.2 × Earth
Mean density: 1.8 × water
Volume: 54 × Earth
Escape velocity: 24.6 km/sec
Axial inclination: 29.6°
Sidereal period of axial
  rotation: 16h 3m
Number of moons: 8
Mean distance from Sun:
  4497 million km
Sidereal period: 164.8 years
Synodic period: 367.5 days
Inclination of orbit: 1.8°
Eccentricity of orbit: 0.009

PLUTO DATA BOX

Diameter: 2250 km
Mass: 0.002 × Earth
Mean density: ~ 2 × water
Volume: 0.01 × Earth
Escape velocity: ~ 1 km/sec
Axial inclination: 117.6°
Sidereal period of axial
  rotation: 6.39 days
  (retrograde)
Number of moons: 1
Mean distance from Sun:
  5900 million km
Sidereal period: 248.5 years
Synodic period: 366.7 days
Inclination of orbit: 17.1°
Eccentricity of orbit: 0.25

PLUTO is a very faint object, appearing as a star of 14th magnitude. It is the smallest of all the planets, with a diameter less than that of our own Moon. Its orbit is so eccentric that from 1979 to 1999 it is closer to the Sun than Neptune, but even at this time it is just a tiny dot among a mass of background stars. The best that amateurs can hope to do is to photograph its movement as it slowly orbits the Sun.

COMETS, METEORS AND ASTEROIDS

COMETS. A bright comet with a glowing tail unfurled across the sky is an awesome sight. Naked-eye comets are rare, but in most years a number of comets come within the range of binoculars and small telescopes. About a dozen new comets are discovered each year, several by amateurs who sweep the sky for just this purpose. Anyone lucky enough to discover a comet has it named after them.

Comets loop around the Sun on highly elliptical orbits that can take anything from a few years to millions of years to complete, depending on the size of the orbit. Those such as Halley's Comet that reappear more frequently than once every 200 years are termed *periodic* (or *short-period*) *comets*; the rest are termed *long-period comets*. Billions of comets are believed to exist in a huge cloud at the rim of the Solar System. Previously unknown comets can approach the Sun at any time from any direction.

The appearance of a bright comet.

When a comet is far from the Sun it is nothing more than a 'dirty snowball' of frozen gas and dust a few kilometres across. But as it approaches the Sun it warms up, releasing gas and dust to form a fuzzy *coma* that can be ten times the diameter of the Earth, yet still mostly transparent. Through small instruments the inner part of the coma appears as a misty patch, circular or elliptical, condensed to a greater or lesser degree towards the central part, now known as the *nucleus*, too small to be visible. In most cases the coma appears to contract as the comet approaches perihelion.

Many comets never develop any further than this, but in some comets sufficient gas and dust flows away from the head to form a noticeable tail, which can stretch for many millions of kilometres. Comet tails are so tenuous that stars can be seen shining undimmed through them. A typical tail stretches for 1° or 2° (two to four Moon diameters), but in exceptional cases can be much longer. Comets are at their brightest – and their tails are longest – near perihelion, which means that they are usually most prominent shortly after sunset or before sunrise when they are inevitably low down in the twilight.

A comet's tail consists of two parts, the *gas tail* and the *dust tail*, which usually diverge as they move away from the comet's head. The gas tail appears bluish in colour, due to the fluorescence of gas molecules, whereas the dust tail is yellowish because it shines by reflecting sunlight. Gas tails point almost directly away from the Sun, whereas dust tails tend to curve more and lag somewhat behind the comet in its orbit. Some comets seem to have an *anti-tail* pointing towards the Sun, but this is a perspective effect, the result of the curving dust tail being seen from an unusual angle. Cometary dust disperses into space and some of it eventually burns up in the Earth's atmosphere to produce meteors (see page 96).

Since comets are so large and diffuse, binoculars and rich-field telescopes are good instruments for observing them. Comets are most easily drawn in negative with a soft pencil on white paper. Record the shape of the coma, its central condensation and any tail. Plot the brightest background stars so that the apparent dimensions of the coma and tail can later be determined from a star atlas.

Comet magnitudes are difficult to estimate. One way is to defocus your binoculars or telescope and compare the image of the comet with that of nearby stars. This technique works well only with brighter, condensed comets. For more diffuse comets, fix the in-focus image of the coma in your mind before defocusing your instrument, and compare this image with the defocused images of nearby stars.

METEORS, popularly known as *shooting stars* or *falling stars*, are streaks of light in the night sky caused by specks of interplanetary dust burning up in the atmosphere at heights of 80 to 100 km. On most clear nights a few meteors are visible to the naked eye every hour. But several times each year the Earth enters dust trails left along the orbits of certain comets. The result is a *meteor shower*, when dozens of meteors per hour may be seen. Whereas non-shower meteors, known as *sporadic* meteors, enter the atmosphere at random, members of a shower all seem to diverge from one small area of sky, the *radiant*, and the shower is named after the radiant's location. For example, the Taurids appear to radiate from Taurus, the η Aquarids from near the star η Aquarii and the Quadrantids from the defunct constellation Quadrans, now part of Boötes. The year's main meteor showers are listed in the table.

Meteors are dust from comets and are usually no larger than a grain of sand. In their structure, though, they are more like a coffee granule, being light, porous and fragile, which is why they vaporize so readily. We do not see the particle itself, only the trail of hot gases it leaves behind as it plunges into the atmosphere at thousands of kilometres per hour.

A meteor shower issues from an area of sky known as its radiant.

MAIN METEOR SHOWERS

Shower	Date of maximum	Radiant RA	Dec.	ZHR (approx.)
Quadrantids *a*	January 3	15.5h	+50°	60
Lyrids	April 22	18.2h	+32°	10
η Aquarids *b*	May 6	22.4h	−01°	35
Perseids	August 12	03.1h	+58°	75
Orionids *b*	October 22	06.5h	+15°	25
Taurids *c*	November 5	03.8h	*c*	10
Leonids *d*	November 17	10.2h	+22°	10
Geminids	December 13	07.5h	+32°	75

a Unusually sharp maximum.
b Associated with Halley's Comet.
c Double radiant: dec. +14° and +22°.
d Major storms every 33 years. Next in 1999?

Meteor observation is one of the easiest branches of amateur astronomy, requiring no special equipment at all. You will need to find a good site with a low horizon. The most comfortable position for scanning the sky is lying in a reclining chair. Remember to wrap up warmly, even for a summer shower such as the Perseids. Face the radiant, but continually survey as much of the sky as possible since the meteors from a shower can appear in any part of the sky.

Aim to watch for at least an hour without a break. Write down the beginning and end times of your observing session, and your observing location. Note the limiting magnitude (faintest stars visible) and percentage cloud cover, as these factors will affect the observed meteor rate. So too will moonlight – there is no point in observing when the Moon is in the sky. Time each meteor to the nearest minute, and estimate its brightness to the nearest magnitude by comparison with the background stars. Note whether it is a shower meteor or a sporadic (in uncertain cases put a question mark). Record any unusual features, such as flares along the path or an explosion at the end, and any colour. Although most meteors vanish in a flash, some leave trains that can last for a minute or more. Give each observation a reliability rating: from A for meteors well seen to C for those glimpsed out of the corner of your eye – of which there will probably be many.

The richness of a shower is indicated by its *zenithal hourly rate* (ZHR), which is the number of meteors that would be seen per hour in clear skies if the radiant were exactly at the zenith; the figures given in the table are only an estimate and vary from year to year, as your observations will help to confirm. The observed rate will be much less than the ZHR if the radiant is low down – for example, the observed rate is half the ZHR for a radiant altitude of 30° and gets rapidly less thereafter. Hence it is best to wait until the radiant is well clear of the horizon before beginning your watch.

The table gives the date of maximum activity, but lesser activity can be seen for several days either side of maximum. In some spread-out showers such as the Perseids and Taurids activity can last for weeks. Graphs of activity plotted from amateur observations reveal information about the structure of meteor orbits.

A really brilliant meteor, of magnitude −3 or more, is known as a *fireball*. Time a fireball to the nearest 5 seconds and, if possible, note its path against the stars (better still, plot it on a chart) so that your observations can be correlated with those of others to establish its trajectory. Some fireballs, not associated with meteor showers, may deliver meteorites to the surface of the Earth. Very rarely, a fireball may be seen as an artificial satellite re-enters the Earth's atmosphere and burns up. There have been occasions when parts of a satellite have survived to land on Earth.

Some amateurs carry out meteor watches with binoculars or telescopes. This allows them to see fainter meteors, but fewer of them because of the restricted field of view. Beginners are recommended to restrict themselves to naked-eye meteor observing.

A Perseid meteor flashes in front of the stars of Aries, photographed on ISO 400 film at f/1.4.

ASTEROIDS, also known as *minor planets*, are small members of the Solar System, debris left over from the formation of the planets. Most of them orbit the Sun in a band between Mars and Jupiter called the *asteroid belt*. The name asteroid means 'starlike', which aptly describes their appearance. Amateur astronomers often overlook the asteroids, despite the fact that the brightest of them, Vesta, reaches magnitude 5.2, brighter than Uranus and visible to the naked eye under ideal conditions; Ceres and Pallas both reach magnitude 6.7 and are easy binocular objects. Ceres is the largest asteroid, with a diameter of approximately 1000 km. Vesta is half the size of Ceres but appears brighter because it is composed of lighter-coloured rock.

In all there are probably several hundred thousand asteroids of all sizes, but only the largest ones are visible in binoculars and small telescopes. It is possible to follow them as they move against the background stars from night to night, and to estimate their brightness by comparison with background stars. Predictions of their positions are given in annual handbooks. If you are uncertain which object is the asteroid, memorize (or draw) the star field and look again the following night.

Very rarely, an asteroid may occult a star. When this is due to happen, predictions of places and times where the event can be seen will be published in astronomy magazines. Amateur timings of the event will assist in deducing the size and shape of the asteroid.

Fragments of asteroids occasionally enter the Earth's atmosphere, either burning up in a brilliant fireball or, in some cases, falling to the Earth's surface as a *meteorite*. These objects are much larger than the dust grains that produce meteors. Several thousand meteorites per year are estimated to reach the surface of the Earth, but most of them go unnoticed because they land in unpopulated areas. A few have caused minor damage to buildings.

STARS

The stars are so far away that they appear as mere points of light through even the largest telescopes, but they are in fact luminous balls of gas similar in nature to the Sun. They are composed predominantly of hydrogen, and produce light and heat by nuclear reactions in their interiors, of which the most important is the conversion of hydrogen into helium. They are born from huge clouds of gas in space called *nebulae*, they live for anything from a few million years to tens of billions of years, depending on their mass, and they die in one of two ways, again depending on their mass.

Astronomers have been able to find out a great deal about the sizes, temperatures and luminosities of individual stars by analysing their light. This is done by splitting the light into a *spectrum* and studying the amount of energy received at different wavelengths. One simple clue to a star's nature is its colour. All stars may appear white at first, but more careful inspection (particularly through binoculars or telescopes) will show that some stars have a noticeably reddish or orange tinge, while others are bluish. The colours are linked to the surface temperature of the stars.

blue giant

Sun

red giant

The Sun compared with larger stars.

Red stars have surface temperatures below about 3500 K; they are the coolest stars of all. (K is the symbol for a degree Kelvin. One degree Kelvin is equivalent to one degree Celsius, but the Kelvin temperature scale begins at absolute zero.) Somewhat hotter stars such as Arcturus appear orange and have temperatures in the range 3500 K to 5000 K, similar to the temperatures of sunspots. Yellow–white stars, of which the Sun is an example, have temperatures between about 5000 K and 6000 K, while pure white stars such as Procyon are hotter still. The hottest stars of all, with temperatures above 11,000 K, appear blue, such as Rigel.

Prominent red stars are Betelgeuse in Orion, Antares in Scorpius (whose name, incidentally, means 'rival of Mars') and Aldebaran in Taurus. Perhaps the most strongly red-coloured of the naked-eye stars is μ Cephei, nicknamed the Garnet Star, whose colour is particularly striking in binoculars. These four are all *giant* or *supergiant* stars, immensely larger and brighter than the Sun. But there are also *red dwarf* stars, far smaller and fainter than the Sun. Although several red dwarfs, such as Proxima Centauri and Barnard's Star, are among the closest stars to the Sun, none of them is bright enough to be visible to the naked eye.

This demonstrates another important factor about stars: their light output or *luminosity*. A large star will have a greater luminosity than a smaller one of the same surface temperature because it is emitting energy from a greater surface area. For example, a red supergiant such as Betelgeuse is about 500 times the diameter of the Sun and over 10,000 times as luminous. At the other end of the scale, Proxima Centauri, a red dwarf, is one-tenth the diameter of the Sun and has only one ten-thousandth of its luminosity.

Another way of expressing the luminosity of a star is by its *absolute magnitude*, the brightness the star would appear to have if it were a standard distance of 10 parsecs from us. On this scale, Betelgeuse has an absolute magnitude of −5.6 (possibly brighter) whereas Proxima Centauri's absolute magnitude is +15.5. The absolute magnitude of the Sun is +4.8. A star's absolute magnitude may be deduced from its spectrum, and provides a powerful way of estimating distances in space.

How is it that stars have such a wide range of sizes? The answer has to do with both the mass of the star

The Sun compared with smaller stars.

(i.e. the amount of gas it contains) and its age. Red dwarfs are small simply because they are much less massive than the Sun, typically one-tenth of a solar mass. The enormous size of giants and supergiants is not such a simple matter – it is a consequence of how these stars evolve. This is perhaps best explained by considering the Sun, which is a typical star.

Our Sun is in stable middle age. It has been shining for nearly 5000 million years and will continue to shine for a similar length of time. As we have seen, had it been less massive it would have been smaller and cooler; and by the same token, had it been more massive it would have been larger and hotter, perhaps like the white and blue–white stars Sirius, Vega and Regulus. Stars in this stable state are termed *main-sequence* stars, whatever their temperature. Eventually, though, they start to exhaust their central store of hydrogen which has fuelled them throughout their life. When this happens, the stars swell up and their outer layers cool and become redder.

In this way the Sun, and stars like it, will eventually evolve into red giants. A star many times more massive than the Sun will become a supergiant. The Sun will die by sloughing off its outer layers to form a so-called *planetary nebula*, leaving behind a tiny *white dwarf* star. Such white dwarfs, the exposed cores of former stars, are about a

---

CLASSIFYING STARS. Astronomers use a somewhat confusing system for classifying stars. First they assign a *spectral type*, in effect an indicator of the star's surface temperature as revealed by its spectrum. There are seven main spectral types, denoted by the letters O, B, A, F, G, K and M, from hottest to coolest, and traditionally remembered by the phrase 'Oh Be A Fine Girl (or Guy) Kiss Me'. The odd sequence of letters is the result of rearranging and simplifying previous systems. Each type can also be subdivided into as many as 10 classes, numbered from 0 to 9.

To distinguish giants from dwarfs, astronomers assign each star a *luminosity class* that runs from I (supergiants) via II and III (bright giants and ordinary giants) to IV (subgiants) V and VI (dwarfs and subdwarfs), and finally VII (white dwarfs). The supergiants are often subdivided into Ia, Iab and Ib to indicate grades of luminosity.

On this classification the Sun is a G2V star (i.e. a dwarf star of intermediate temperature). The blue supergiant Rigel is classified as B8Ia, and the red supergiant Betelgeuse as M2Iab. Sirius, a main-sequence star somewhat hotter than the Sun, is A1V, and the red dwarf Proxima Centauri is M5.5V.

hundredth of the diameter of the present-day Sun and are exceptionally dense: a spoonful of white dwarf material would weigh many tonnes. Sirius, the brightest star in the sky, has a white dwarf as a companion star in a 50 year orbit around it. This white dwarf has an apparent magnitude of 8.3, but is drowned by the brilliance of Sirius and so is difficult to see without a large telescope. The easiest white dwarf to see is a magnitude 9.5 companion of the star o^2 Eridani.

The most massive stars, those that become supergiants, die in an altogether more spectacular way. They have the shortest lives, consuming their central stores of hydrogen in only a few million years, and then erupt catastrophically as *supernovae*, spouting their remains into space at high speed to form *supernova remnants*. Supernovae leave behind objects that are even more extreme than white dwarfs. In some cases the core of the supernova is crushed to become a superdense *neutron star* containing as much mass as the Sun packed into a ball only 20 km across.

For a neutron star of more than about three solar masses, the final result is the most bizarre of all. Theory predicts that it will continue to shrink under the inward pull of its own gravity until it becomes a *black hole* with a gravitational field so strong that even its own light can no longer escape. Needless to say, such objects are beyond the capacity of amateur observation, although one putative black hole, Cygnus X-1, has a ninth-magnitude companion star that is visible in moderate telescopes.

DOUBLE STARS. A large proportion of stars have one or more companion stars, and small instruments will show many of these groupings. In some cases the stars are not truly associated but simply happen to lie in the same line of sight – these are termed *optical doubles*. But the great majority of double stars are true *binaries*, in which the two stars are orbiting each other. There are also triple, quadruple and even larger families of stars, evidently formed together from the same cloud of gas. The closest star to the Sun, *α* Centauri, is in fact a triple star system, but it is visible only from latitudes south of 30°N. It turns out that single stars such as the Sun are in the minority.

One of the delights of observing double stars is the contrasting colours of many pairs, which are often more noticeable than the colours of individual stars. The best example, and the double star that most observers cut their teeth on, is *β* Cygni, also known as Albireo. It consists of an

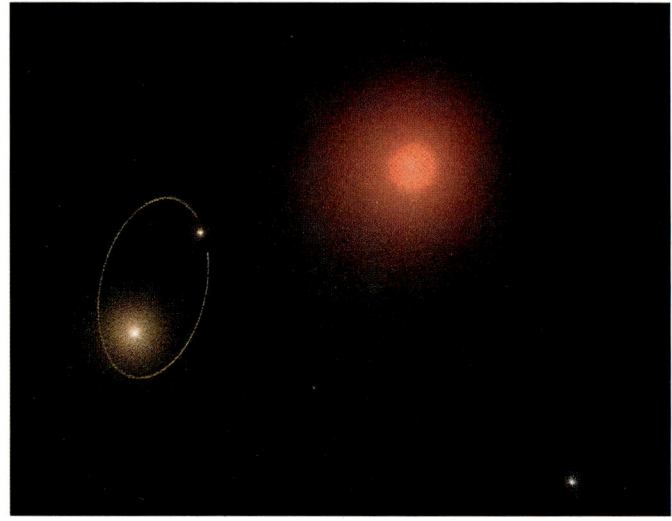

Double stars: many doubles are genuine binaries, as on the left, but in other cases the two stars lie in the same line of sight but at widely differing distances from the Earth, as on the right.

orange giant star (spectral type K) of magnitude 3.1, and a dwarf of magnitude 5.1 and spectral type B that appears blue–green by contrast with the primary. The smallest of telescopes will divide, or 'split', this pair. A similar pair with contrasting colours but smaller separation is γ Andromedae.

In other cases the stars can appear virtually identical to each other in colour, and sometimes also in brightness. Look, for example, at 61 Cygni, a pair of orange dwarf stars of magnitudes 5 and 6, easily split in small telescopes, or at γ Leonis, a much tighter pair of orange giants that require high magnification to be split. Star colours, incidentally, can be more noticeable if the telescope is slightly defocused or if you tap the telescope to make the images vibrate.

Of all the doubles in the sky, the widest lies in the handle of the Plough and can be split by keen eyesight. The two stars are Mizar and Alcor, of magnitudes 2 and 4, both blue–white main-sequence stars. They are too far apart to form a genuine binary, but simply move through space together. Small telescopes reveal another fourth-magnitude companion much closer to Mizar, and these two form a genuine binary. Another wide double, just divisible by keen eyesight but easy enough in binoculars, is ε Lyrae near the brilliant star Vega. This star is in fact a fascinating multiple

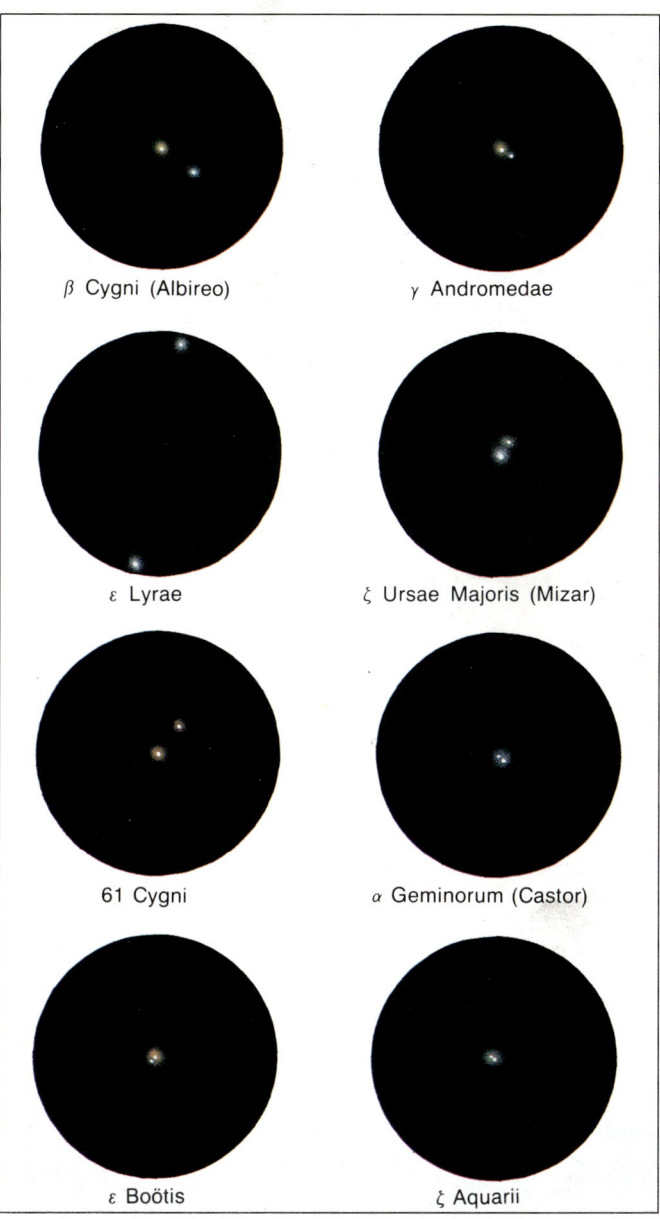

Some celebrated double stars as seen through a small telescope.

star: as 75 mm (3 inch) telescopes with high magnification will show, each star of the pair is itself double. All these stars are linked by gravity, thus forming a rare quadruple unit.

Small telescopes can split wide double stars, but larger apertures (which have better resolution) are needed to split close doubles. The resolutions of various telescope apertures are given in the table on page 31. However, this table is not an infallible guide to the closeness of double stars that can be split, since the brightness of the components also affects the outcome. A pair of unequal brightness will be much more difficult to split in a given aperture than a near-identical pair. This is why, for example, the faint white dwarf companion of Sirius is so difficult to see. An additional factor is the unsteadiness of the atmosphere (the seeing). When the seeing is poor, the resolution of your telescope will be severely impaired.

Some pairs are too close together to be split in even the largest telescopes. In such cases we know that a star is double only because a double spectrum is apparent when its light is analysed in a spectroscope. Such stars, called *spectroscopic binaries*, can have orbital periods as short as a few days or even hours. In some cases, the orbit is presented to us virtually edge-on, and the two stars actually pass in front of each other as seen from Earth – these are called *eclipsing binaries* (see page 113).

Mizar, its close fourth-magnitude companion and Alcor are all spectroscopic binaries, making this a highly complex system. Another bright multiple star containing spectroscopic binaries is Castor (α Geminorum). To the naked eye, Castor appears as a blue–white star of magnitude 1.6. Small telescopes split it into a dazzling pair of second- and third-magnitude stars 4 arcsec apart that orbit each other in approximately 500 years. A third component, a ninth-magnitude red dwarf, lies 70 arcsec away. All three of these stars are spectroscopic binaries, so that Castor is in fact a set of stellar sextuplets.

In many cases the orbital periods of double stars are so long – many centuries or even millennia – that positional changes of the two stars are undetectable to the eye. However, some binaries have much shorter orbits and can change noticeably in appearance from year to year. For example, ξ Ursae Majoris has a period of 60 years. In 1985 the fainter star of the pair lay due east of its companion, but by the year 2000 it will lie due west, having traversed 180° in just 15 years (see diagram). Other stars with rapid orbital motions that can be followed in small telescopes are ζ Herculis (period 34.5 years), ι Leonis (period 192 years), 70 Ophiuchi (period 88 years) and γ Virginis (period 171 years). Optical doubles, although physically unrelated, can change their orientation and separation over time because of each star's *proper motion* – its own movement through space.

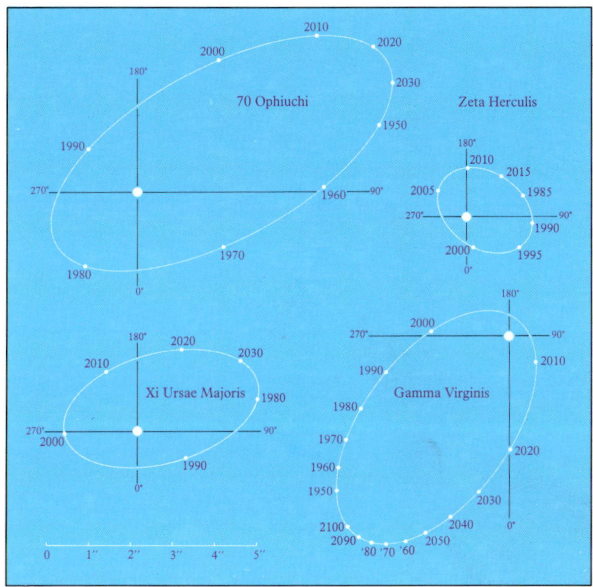

Orbits of four fast-moving binary stars.

SEPARATIONS AND POSITION ANGLES. The relative positions of the components of a double star are described by their *separation* and *position angle* (PA). Separation is measured in arc seconds. Position angle is the orientation of the fainter star with respect to the brighter one, measured counterclockwise on the sky in degrees from north (0°) via east (90°), south (180°) and west (270°). Note that west is the direction in which stars drift through the field of view. To estimate the position angle of a pair, allow them to drift through the telescope's field of view – this establishes the east–west line (90° to 270°) in the sky. Alternatively, observe the pair when it is near the meridian, so that 0° lies towards the bottom of the field of view in an inverting telescope, and east is to the right. With practice, it is possible to estimate the PA of a pair to within 10° by eye alone. Some advanced amateurs use micrometers to make precise measurements of double stars from which their orbits can be calculated.

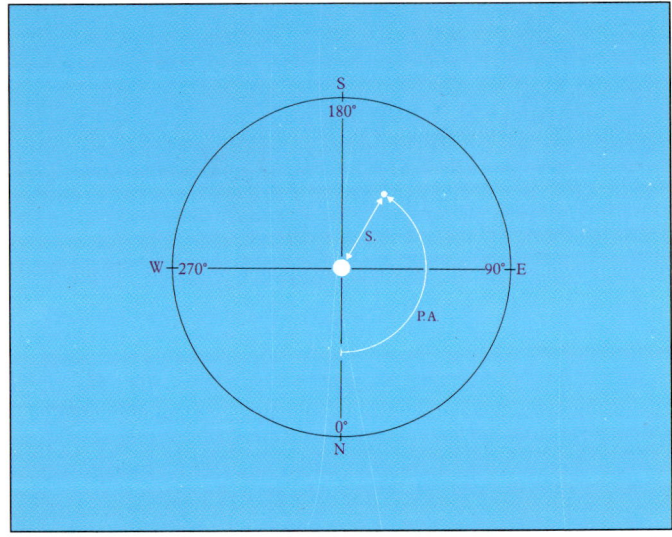

The position angle (PA) of a double star is measured in degrees from north through east. Separation (S) is measured in seconds of arc.

VARIABLE STARS. Many stars vary in brightness over periods ranging from less than a day to several years. These variations can be followed by amateurs using small instruments, or even with the naked eye. There are so many variable stars that professional astronomers cannot monitor them all, and amateur observations provide valuable information that helps in the understanding of stars. In most cases the variations are caused by physical changes in the stars, but in nearly 20% of known variable stars the changes are caused by eclipses of close doubles (see page 108).

Among the most interesting variables are the red giants and supergiants, which include some of the brightest naked-eye stars. Such stars have swollen up in size at the end of their life and have as a result become unstable, changing erratically in size and brightness. Most abundant of all types of known variable are the *long-period variables*, also known as *Mira stars* after their prototype, Mira (o Ceti), which was the first variable star to be noticed. Mira itself is an M-type giant that varies between about third and ninth magnitudes over a period of about 11 months, although at times it can become as bright as second magnitude. When at its brightest it is a clear naked-eye object, and can be followed in binoculars during its rise and fall with the help of the comparison chart shown here. Other Mira stars have periods from three months to three years, and amplitudes up to 10 magnitudes.

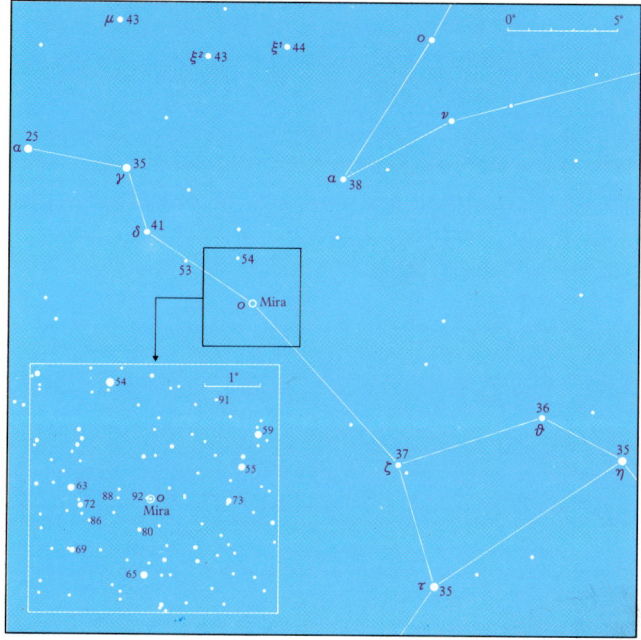

Comparison/finding chart for Mira. (The magnitudes of comparison stars are given without decimal points, e.g. 54 not 5.4, to avoid confusion with stars.)

Semiregular variables are similar to the Miras, but with much less well-defined periods, and amplitudes of only one or two magnitudes. Prominent examples are Betelgeuse (α Orionis), which ranges between about magnitudes 0.4 and 1.3; α Herculis, range 2.7 to 4.0; and μ Cephei, range 3.4 to 5.1. All three are M-type supergiants. There are also *irregular variables*, with no detectable period, of which the M-type giant β Pegasi is an example, ranging between 2.3 and 2.7.

Perhaps the most important type of variable star is the less common class known as *Cepheid variables*, after their prototype δ Cephei. These are supergiants and bright giants of spectral classes F to K (i.e. white to orange in colour). They are going through a specific stage of their evolution during which they pulsate in size, but unlike the red stars mentioned above the pulsations of each Cepheid are highly regular, with periods from about one day to four months. What makes them of particular importance is that a Cepheid's period of pulsation is directly linked to its absolute magnitude, the slowest pulsators being the most luminous. This gives astronomers a convenient way of measuring the distance to a Cepheid: simply measure its period, which reveals its luminosity, and compare this with its apparent brightness. Cepheids have proved to be important distance indicators in our Galaxy, and also for deducing the distances of nearby galaxies.

δ Cephei itself ranges between magnitudes 3.5 and 4.4 every 5.37 days. These changes can easily be followed by the naked eye or binoculars with the help of the chart given here.

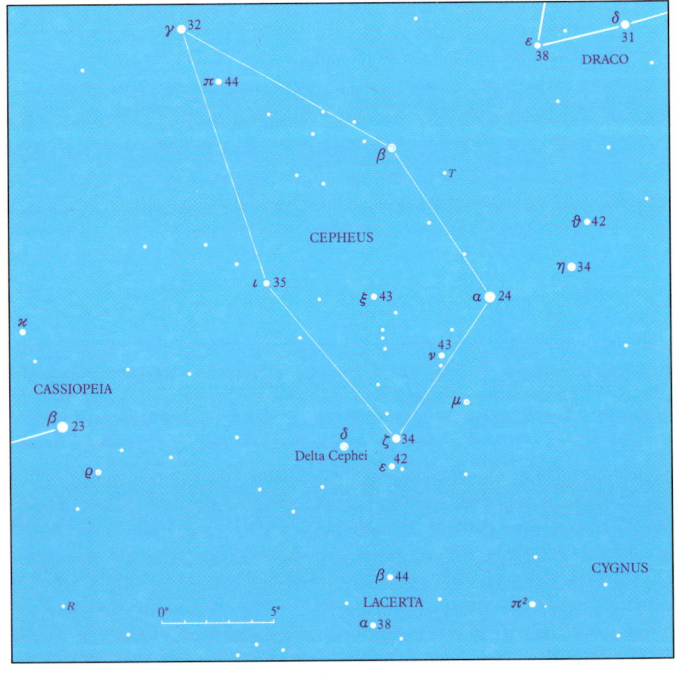

Comparison chart for δ Cephei. (The magnitudes of comparison stars are given without decimal points.)

By estimating the magnitude of δ Cephei against the comparison stars on the chart it should be possible to produce your own graph of its light variations (the *light curve*) like the one shown. Two other bright Cepheid variables to look at are η Aquilae (magnitude range 3.5 to 4.4, period 7.18 days) and ζ Geminorum (3.6 to 4.2, 10.15 days).

Completely different in nature are *eclipsing binaries*, in which the variations are caused not by physical changes in the stars themselves but by eclipses (either partial or total) in

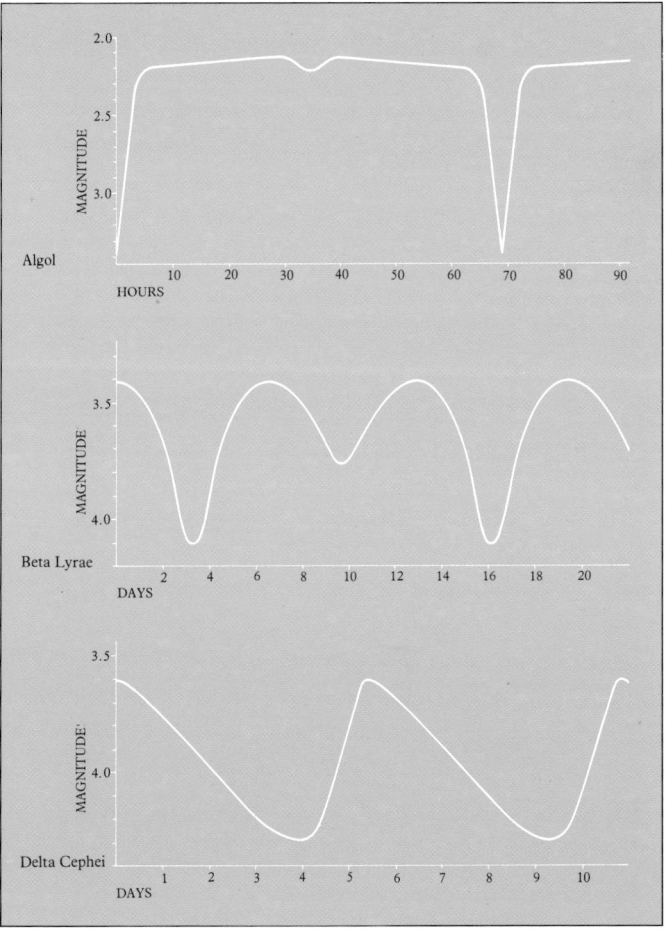

Light curves of the variable stars Algol, β Lyrae and δ Cephei.

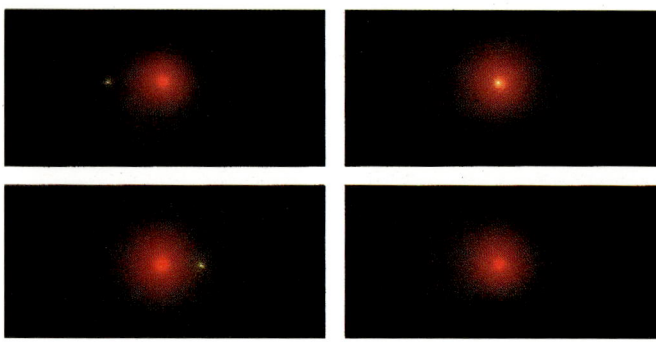

An eclipsing binary changes in apparent brightness as the stars orbit each other. When both stars are visible (above and below left) the star is at maximum brightness. When the stars eclipse each other (above and below right) the total brightness drops.

close binary stars. The prototype eclipsing binary is Algol (β Persei), which fades every 2.87 days from magnitude 2.1 to 3.4, taking five hours to sink to minimum and another five hours to regain its former brightness. Algol consists of a blue–white main-sequence star and a yellow subgiant. The fading occurs when the larger but fainter subgiant partially eclipses its brighter companion. A secondary eclipse occurs

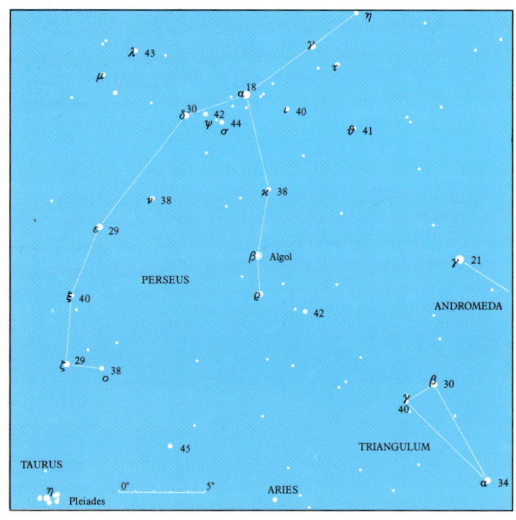

Comparison chart for Algol. (The magnitudes of comparison stars are given without decimal points.)

half an orbit later when the blue–white star passes in front of the yellow star, but the dip in brightness is too small to be noticeable to the naked eye. Hence the brightness of Algol is effectively constant outside the main eclipse. You can follow the changes of Algol by using the comparison chart here. A similar star is λ Tauri, which ranges from magnitude 3.4 to 3.9 every 3.95 days.

An extreme form of eclipsing binary is β Lyrae, in which the two stars are so close that they have been drawn into egg shapes by each other's gravity. β Lyrae varies from magnitude 3.3 to 4.3 every 12.9 days, the brightness continually changing as the distorted stars move around their orbits.

Totally unpredictable are *novae* and *supernovae*. They are the volcanoes among variable stars. Novae were once thought to be new stars bursting into view (the word 'nova' is Latin for 'new'), but we now know that they occur in close binaries, one member of which is a white dwarf. Gas spills over from the companion star onto the surface of the white

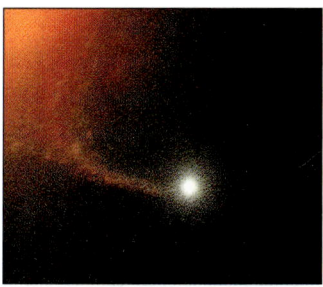

How a nova occurs: gas from a companion star flows onto a white dwarf (right and below), igniting in a nuclear eruption (below right) that throws off a shell of gas.

dwarf, where it occasionally ignites in a nuclear explosion. This causes the binary to increase in brightness by anything from a thousand to millions of times, bringing a formerly anonymous star into sudden prominence. The brightest nova of recent years was Nova Cygni 1975, which brightened by at least 19 magnitudes to reach second magnitude for two days.

Such naked-eye novae are rare, but each year several novae become visible to the observer with binoculars. A typical nova rises to maximum in a few days, remains near its peak for no more than a week and then declines to its original level, often erratically, taking several years. Amateurs provide valuable help in monitoring the activity of novae. Certain novae have been seen to re-erupt, most notably RS Ophiuchi, which underwent outbursts in 1898, 1933, 1958, 1967 and 1985. It is probable that, given time, all novae will recur. Many novae are discovered by amateurs as a result of deliberate searches, now mostly photographic, although some continue to use traditional visual methods.

Even more astounding are supernovae, which brighten by more than 100 million times (20 magnitudes), and in extreme cases can equal the entire light output of a galaxy. Supernovae are the explosive deaths of stars, either white dwarfs in binary systems in the case of *Type I supernovae*, or supergiants in the case of *Type II supernovae*. No supernova has been seen in our Galaxy since 1604, although undoubtedly some have occurred and been masked by dust clouds. In 1987 a supernova erupted in the neighbouring Large Magellanic Cloud, reaching third magnitude.

Amateurs conduct regular searches of other galaxies for supernovae, although large telescopes are needed because the galaxies are distant and the supernovae faint. Statistics from such searches suggest that two or three supernovae should occur each century in our Galaxy. We are long overdue for one – it could come at any time.

You can estimate the brightness of a variable star by comparing it with nearby stars of known magnitude. With practice, it is possible to make estimates accurate to about a tenth of a magnitude. You may find it helpful to defocus the images of stars into disks when comparing their brightnesses; this technique is particularly useful for bright or strongly coloured stars. In the *step method* the observer estimates the difference in brightness between the variable and a

---

NAMING VARIABLE STARS. Variable stars that do not already have a Greek letter or Flamsteed number are designated, according to a specific but somewhat complicated system, by one or two capital letters – for example R Leporis and BU Tauri. This allows for a total of 334 variables per constellation, but even this is insufficient in some constellations. The identification system then continues indefinitely with V335 and subsequent numbers prefixed by the letter V. Novae are also identified by this system.

---

comparison star in steps of 0.1 magnitude. The process is repeated with a series of comparison stars, and the results are averaged. Wherever possible the comparison stars should be at a similar altitude to the variable star to prevent errors caused by atmospheric dimming.

A simpler alternative is the *fractional method*, in which the observer estimates where the brightness of the variable star (V) lies in relation to comparison stars, one brighter (star A) and one fainter (star B). For example, if the brightness of the variable appears to be exactly halfway between those of the comparison stars, the observation is written in the form A1V1B. If the variable seems to be two-thirds of the way in brightness from A to B, the observation is written A2V1B; and so forth. The observation can be repeated with another pair of comparison stars. Results should be rounded to the nearest tenth of a magnitude.

Variables such as δ Cephei should be observed on every clear night, whereas Algol-type binaries need be observed only when an eclipse is in progress (predictions are carried in astronomy magazines). For Algol itself brightness estimates should be made every half-hour for several hours either side of mid-eclipse. For stars such as Mira, Betelgeuse and Antares, estimates once or twice a week should be sufficient. Glance also at γ Cassiopeiae from time to time, since it can undergo sudden unexpected changes. Above all, try not to be biased by your previous estimates or by assumptions about how bright you expect the star to be.

Betelgeuse, the red supergiant in Orion, is a variable star whose slow changes in brightness can be followed by checking every week or so.

CLUSTERS, NEBULAE AND GALAXIES

STAR CLUSTERS are groups of stars bound together by gravity. There are two main kinds: *open clusters*, which lie in the spiral arms of our Galaxy, and *globular clusters*, which lie in a spherical halo around it. Open clusters can consist of anything from a dozen or so to several hundred stars, all relatively young, whereas globular clusters contain hundreds of thousands of very old stars.

Open clusters are irregular in shape and vary in appearance from quite tightly packed to loose and scattered. The most famous open cluster – and the easiest to see – is the Pleiades in the constellation Taurus. The Pleiades cluster, designated M45, is commonly called the Seven Sisters because seven stars can be seen with the naked eye under good conditions (although with excellent eyesight and clear skies, more are visible). Binoculars reveal dozens more, and probably several hundred stars belong to the cluster. The brightest star in the Pleiades is Alcyone, a blue giant of magnitude 2.9.

All the stars in an open cluster were born relatively recently from the same cloud of gas and dust. In the Pleiades, which is considered to be no more than about 50 million years old (very young by stellar standards), some traces of the original cloud can be seen on long-exposure photographs. Another famous open cluster is the Hyades, also in Taurus, consisting of about a dozen naked-eye stars arranged in a V-shape. Dozens more are visible in binoculars. The Hyades cluster is about ten times older than the Pleiades and is more widely scattered. Most open clusters eventually disperse, although some have lasted for several thousand million years. The red giant Aldebaran is not a member of the Hyades, but a superimposed foreground star.

The Pleiades star cluster is an ideal object for binoculars.

Since open clusters are usually large, they are well suited to observation by low-power instruments with wide fields of view. The Pleiades, for example, covers 1° of sky (twice the diameter of the Moon) while the Hyades is spread over 5° so that only binoculars can take in the whole of it at once. Another large cluster lies in the constellation Cancer. It is variously called Praesepe, the Beehive, the Manger or simply M44. It is visible to the naked eye as a misty patch on clear nights covering about 1.5°. Binoculars show a scattering of dozens of stars of sixth magnitude and fainter. The most scattered open cluster of all is a sprinkling of fifth- and sixth-magnitude stars that straggles for 10° across the constellation Coma Berenices.

Many clusters appear diffuse, like comets, when viewed through binoculars and small telescopes. Look, for example, at the group of three open clusters in Auriga, M36, M37 and M38, all of which just fit into the same field of view in wide-angle binoculars. A telescope will show the different character of each cluster: M36 – the middle one of the trio – is the smallest, M37 is the largest and brightest but looks hazy because its stars are faint, while M38 is the loosest.

A characteristic of many open clusters is the presence of one or more red giant stars near the centre, as in M37. A particularly beautiful example is the Wild Duck Cluster (M11), in Scutum, so named because its brightest stars form a fan shape resembling ducks in flight. At the apex of the fan lies an eighth-magnitude red giant. A similar configuration can be seen in M50 in Monoceros and M52 in Cassiopeia.

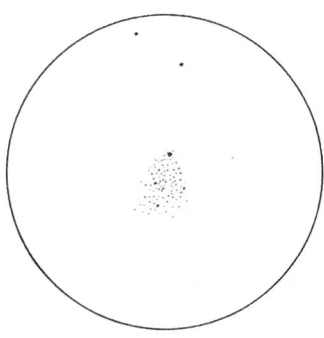

M11, the Wild Duck Cluster, 150mm reflector, ×48.

Several clusters have unusual shapes. In many the stars appear to be arranged in chains, sometimes forming a cross. The outline of NGC 2264 in Monoceros has been compared to a Christmas tree, while M6 in Scorpius is called the Butterfly Cluster. You should look out for any unusual configurations when observing open clusters.

Near M6 is one of the most spectacular open clusters of all: M7, visible to the naked eye as a bright knot in the Milky Way, 1° across and easily resolved by binoculars into its component stars. Both M6 and M7 are too far south to be well seen by many observers in the northern USA and Europe, but an excellent substitute is the Double Cluster in

NAMING CLUSTERS, NEBULAE AND GALAXIES. These are known collectively as *deep-sky objects*, and are usually referred to by numbers prefixed by the letters M or NGC. The 'M' prefix is a testament to the work of Charles Messier, an eighteenth-century French astronomer who compiled a catalogue of over 100 deep-sky objects. The NGC designation stands for *New General Catalogue*, a major listing of over 7800 objects, subsequently increased to more than 13,000 by two *Index Catalogues* (IC).

Perseus, two subtly different groupings easily found in binoculars. The brighter and richer of the pair is NGC 869. Its companion, NGC 884, is apparently older for it contains several red giants, which NGC 869 does not.

Globular clusters, as their name implies, are ball-shaped. They are usually spherical but some have a noticeably elliptical outline. Globulars are larger than open clusters, some being well over 100 light years in diameter, but they contain many more stars than open clusters do and thus appear much more densely packed, usually more concentrated towards the centre. In the largest globulars there are estimated to be over a million stars, and the stars in globular clusters are among the oldest in our Galaxy.

More than a hundred globular clusters are known, but only a few dozen are easily visible through amateur instruments. In binoculars and small telescopes they appear smudgy, very much like comets, but larger apertures and a high magnification can resolve several of them. The largest and brightest globulars, ω Centauri and 47 Tucanae, lie deep in the southern hemisphere.

The best globular cluster in the northern sky is M13 in Hercules, just visible to the naked eye in clear skies as a sixth-magnitude misty spot half the apparent diameter of the Moon. M13 is estimated to contain several hundred thousand stars in a volume about 100 light years in diameter, and is about 25,000 light years away. A 100 mm (4 inch) telescope will show some of the brightest red giants in this cluster.

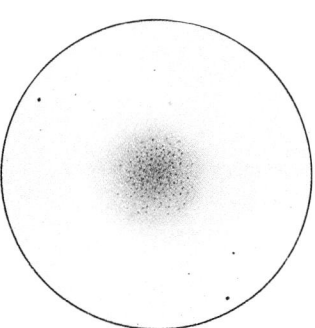

M13 globular cluster in Hercules, 150 mm reflector, ×48.

Another good northern globular, rivalling M13, is M5 in Serpens. It is somewhat elliptical, and has chains of stars radiating across its outer regions. The closest of the bright globulars is M4 in Scorpius, about 7000 light years away, but it is difficult to spot in poor skies because its stars are spread over nearly half a degree. About half as far away again is M22 in Sagittarius, a large fifth-magnitude globular that is a superb sight in all apertures.

NEBULAE are clouds of gas and dust in the spiral arms of our Galaxy. They are elusive objects to observe because they are naturally diffuse, but the most famous of them, the Orion Nebula (M42), can be seen with the naked eye on clear nights. In common with many nebulae the Orion Nebula has a rather ethereal greenish hue, more obvious in binoculars and small telescopes. This colour is due to ionized oxygen and is more prominent to the eye than is the reddish light from hydrogen, which is the main component of all nebulae. Photographs, though, show the reddish colour of nebulae strongly, as most photographic emulsions are more sensitive to red light.

The shape of the Orion Nebula has been variously compared to a flower, a bird and even a giant ghostly bat. Suffice it to say that this great cloud, spread over more than 1°, provides an endless source of interest for users of all

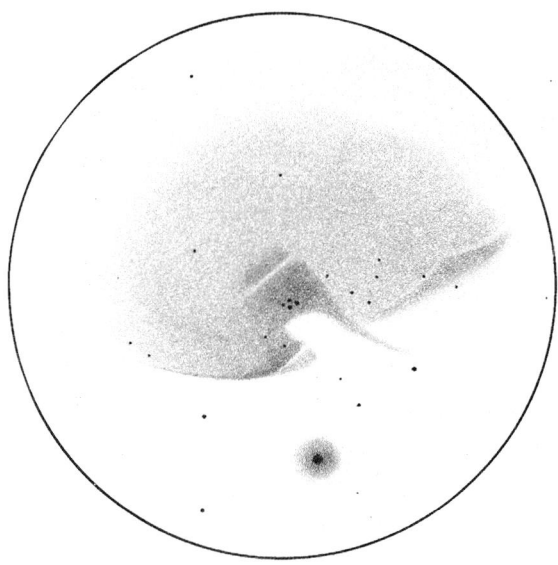

M42, the Orion Nebula, 200mm catadioptric, ×80.

apertures, seeming to display more detail in its diaphanous veils each time it is observed. It lies some 1600 light years away and is about 30 light years in diameter. Next to it is M43, which long-exposure photographs show to be part of the same huge cloud, separated from M42 by darker material.

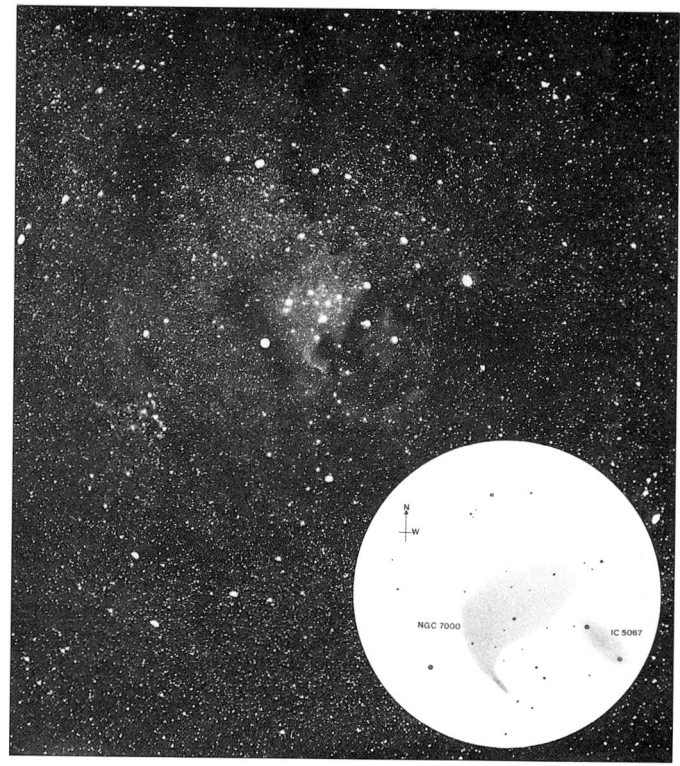

The North America Nebula, NGC 7000, photographed with a 1 hour exposure through an 85mm lens at f/2 on ISO 200 film, and (inset) as seen through an 8×50 finder telescope with a nebula filter.

All bright nebulae contain young stars, and this indicates a two-way relationship. Stars are born from diffuse nebulae, and the light from the new-born stars then makes the nebula glow. In the Orion Nebula the illuminating stars are not hard to find: near its centre lies a knot of four stars, visible in small telescopes, called the Trapezium (in addition, two fainter stars can be seen in apertures of 100 mm/4 inches or more). They lie at the end of a dark inlet called the Fish Mouth. Astrophysicists have found that a whole cluster of stars is coming into being inside the Orion Nebula.

Another beautiful example of a gas cloud and cluster is the Lagoon Nebula (M8) in Sagittarius, 1° long and with a backbone of stars. The Lagoon is an easy object in binoculars, and is just visible to the naked eye. In some cases the cluster is easier to see than the surrounding nebula, as in the case of M16 in Serpens. Here the cluster looks hazy in small instruments because it is embedded in the Eagle Nebula, but the nebula shows up well only on long-exposure photographs. NGC 2244 in Monoceros is an oblong cluster, visible in binoculars, surrounded by a large, pale ring known as the Rosette Nebula. The Rosette can be seen in binoculars under clear, dark skies but its full beauty is brought out only on photographs.

Sometimes the illuminating star is more difficult to find, as in the North America Nebula (NGC 7000) in Cygnus. This celebrated object, shaped like the continent after which it is named, spans more than 1.5° and can be seen in binoculars in dark skies. There are plenty of stars within the nebula, although which of them is responsible for making it shine remains something of a puzzle.

If there are no stars within a nebula then the cloud remains dark and can be seen only in silhouette against a lighter background. For example, darker obscuring material is responsible for the 'Gulf of Mexico' indentation in the North America Nebula, as well as for the Fish Mouth in the Orion Nebula. A large dark cloud blots out part of the Milky Way in Cygnus, forming the so-called Cygnus Rift. Perhaps the most unusual dark nebula is the Horsehead, shaped like a chess knight, that lies in front of bright nebulosity south of the star ζ Orionis. Unfortunately, although the Horsehead looks spectacular on photographs it is notoriously difficult to see in telescopes.

Other types of nebula represent not the birthplaces of stars but their deathbeds. Stars such as the Sun slough off their outer shells at the end of their life to form *planetary nebulae*, so named not because they are related to planets but because when they were first observed their greenish-blue disks were thought to resemble the planet Uranus. The most conspicuous planetary nebula is the Dumbbell Nebula (M27) in Vulpecula. It appears as a round, misty patch in binoculars, but telescopes reveal the hour-glass shape that gives it its

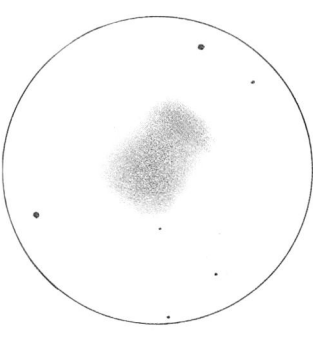

M27, the Dumbbell Nebula, 150 mm reflector, ×48.

name. While the Dumbbell can be found in average skies, another large planetary, the Helix (NGC 7293) in Aquarius, requires dark, clear skies to be seen because of its low surface brightness.

Several other planetaries are within the reach of small telescopes, among them NGC 7662 in Andromeda, visible as an elliptical greenish disk with magnifications of 100 times or so. Another good one is NGC 6826 in Cygnus, less than 1° from the sixth-magnitude double star 16 Cygni. NGC 6826 is known as the 'blinking planetary' because looking alternately towards and away from it creates the impression that it is blinking on and off. This optical illusion is apparent only in apertures greater than about 100 mm (4 inches), though.

The archetypal planetary nebula is M57 in Lyra, popularly known as the Ring Nebula. Although it looks like a ninth-magnitude fuzzy star in small telescopes, an aperture of 100 mm (4 inches) or more will show the famous 'smoke ring' shape, slightly elliptical in outline, that gives the nebula its name.

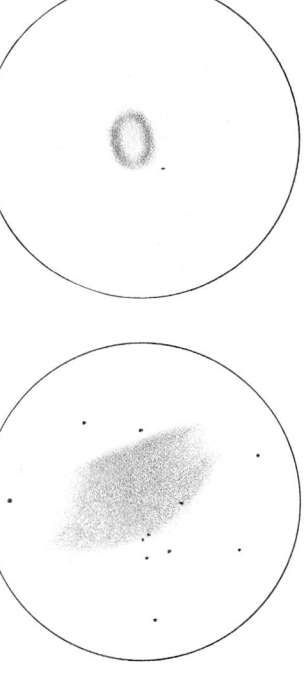

As we saw on page 105, stars much more massive than the Sun erupt as supernovae at the end of their life. One such supernova was seen in AD 1054 in the constellation Taurus, and the remains of that eruption can still be seen as the Crab Nebula (M1). The Crab is one of the most famous objects in the sky yet it is not particularly easy to spot. Look for an elliptical misty patch just over 1° from ζ Tauri. The Crab Nebula is about six times the apparent size of Jupiter, and beginners often overlook it because it is larger and fainter than they are

Above: M57, the Ring Nebula, 216mm reflector, ×135.
Below: M1, the Crab Nebula, 406mm reflector, ×110.

expecting. Only large apertures will reveal the bright filaments of gas that give the nebula its crab-like appearance.

Remains of another supernova are to be found in Cygnus, in the form of the Veil Nebula. At best, all that can be seen in amateur telescopes is a few twisted strands of gas (unless special filters are used), but the brightest section, NGC 6992, can just be glimpsed in binoculars under ideal conditions.

FILTERS. Many nebulae are easier to see with the help of special *nebula filters* that can be fitted over your telescope's eyepiece. They let through only the narrow range of wavelengths that nebulae emit, and so make the nebula stand out strongly against a dark background. They have names such as UHC (ultra-high contrast) and Oxygen III (referring to the greenish light from oxygen). Light pollution rejection (LPR) filters are designed to cut out unwanted light from streetlights, and are suitable for photography or visual observation of deep-sky objects. These drawings are of NGC 6992 as seen (*a*) without and (*b*) with a nebula filter.

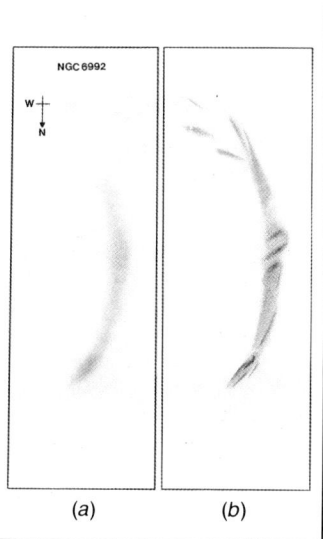

(*a*)   (*b*)

GALAXIES. Our Sun is a member of the Galaxy, a system of perhaps a quarter of a million stars arranged in the shape of a flattened spiral with a central bulge, or nucleus. Most of its stars are concentrated in the luminous band of the Milky Way, which marks the plane of the Galaxy. The stars of the Milky Way are densest in the constellation Sagittarius, which is where the centre of our Galaxy lies, about 30,000 light years away. On a dark, clear night the star fields of the Milky Way offer breathtaking views through binoculars and wide-field telescopes. The Sagittarius star fields lie south of the celestial equator, so are not well seen by far-northern observers. Our Galaxy has two small companions, the Large Magellanic Cloud and the Small Magellanic Cloud, which appear in the night sky like detached portions of the Milky Way, but they lie so far south that they can be seen only from close to the equator or from the southern hemisphere.

There are countless other galaxies throughout space, the nearest and brightest of which are within the range of small telescopes. One of them can be glimpsed with the naked eye – the great *spiral galaxy* M31 in Andromeda. Although 2.4 million light years away it is visible as an elongated blur on dark nights, and binoculars bring it out unmistakably. M31 shows us what our Galaxy would look like if we could see it from the outside: a whirlpool of stars over 100,000 light years in diameter. Depending on the clarity of the sky and the instrument used, M31 can be traced outwards from its bright nucleus for 1° or more in each direction along its longest axis. Small telescopes will also show traces of the galaxy's spiral arms wrapped tightly around the nucleus, but you will not be able to resolve individual stars. M31 looks elliptical because its plane makes an angle of less than 20° with our line of sight.

Another nearby spiral galaxy is M33 in Triangulum, but it has a low surface brightness and is much more difficult to see than M31. On a dark night this object can be picked out in

M31, the great spiral galaxy in Andromeda, photographed through a 300 mm telephoto lens at f/4.5, exposure 45 minutes, ISO 200 film.

binoculars as a rounded misty patch about the same apparent size as the Moon. It is 2.6 million light years away.

Not all galaxies have the same shape as our Galaxy and M31. Some have a bar of stars across their centre, and hence are known as *barred spiral galaxies*. They are very distinctive on photographs, but none show up well in small telescopes. The third main type are the *elliptical galaxies*. They include the largest known galaxies, some with ten times as many stars as our Milky Way, and – at the other end of the scale – the dwarf ellipticals, which by their very nature are faint and difficult to spot. M31 has two small elliptical companions. One of them (M32) is visible in small telescopes half a degree south of M31's core. The other (NGC 205) is fainter and more difficult to see, a degree from the heart of M31. Some small galaxies are classified as *irregular*, of which the Magellanic Clouds are examples.

Our Galaxy, M31 and M33 are the three largest members of a cluster of about thirty galaxies called the Local Group. Many other galaxies belong to clusters such as the famous Virgo Cluster (which actually spills northwards into neighbouring Coma Berenices). The Virgo Cluster contains over 2000 galaxies, some of which are bright enough to see in amateur telescopes even though they are 50 million light years away. At the heart of the Virgo Cluster lies M87, a giant elliptical visible in apertures of 100 mm (4 inches) and above as a ninth-magnitude rounded fuzz.

What is perhaps the best-known galaxy in Virgo is not a member of the cluster at all, but a foreground object: M104, known as the Sombrero Galaxy because it looks a bit like a Mexican hat. This eighth-magnitude galaxy is seen nearly edge-on and can be found in apertures of 100 mm (4 inches). It is crossed by a dark lane of dust that normally needs at least 150 mm (6 inches) to be detected. Another good example of the presence of dust in galaxies is M64 in Coma Berenices, known as the Black Eye Galaxy because of a dust lane seen in silhouette near its nucleus; the galaxy itself, of magnitude 8.5, can be seen in a 100 mm (4 inch) telescope but the black eye feature may need larger apertures.

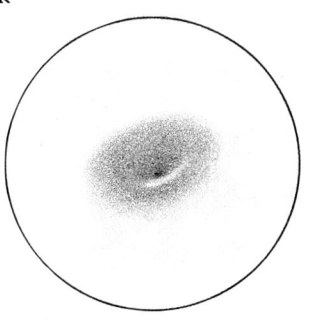

M64, *the Black Eye Galaxy, 406mm reflector,* ×203.

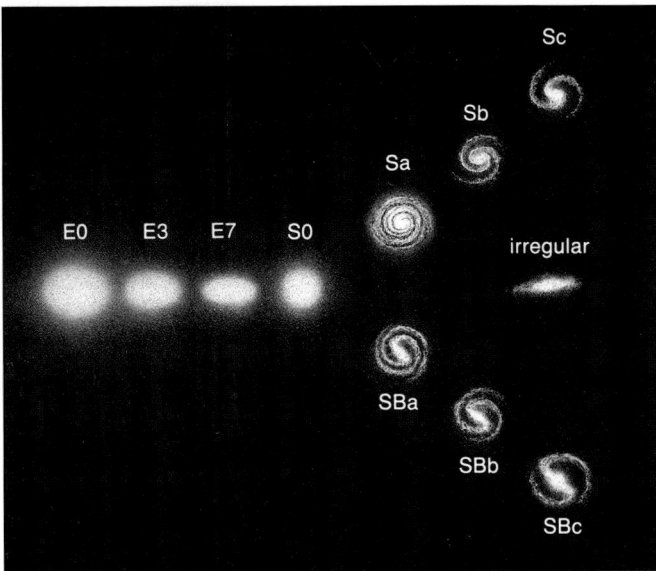

The three main types of galaxy are shown in this diagram. At the left are elliptical galaxies, classified from E0 to E7 depending on their degree of ellipticity. Ordinary spiral galaxies (type S) and barred spirals (type SB) are subdivided depending on how tightly their arms are wound around their centre. Galaxies of type S0 are spirals without arms. A galaxy's type is fixed at birth; galaxies do not evolve from one type into another, unless two galaxies meet and merge.

Unusual things seem to be happening in some galaxies. In M87, mentioned above, a colossal explosion seems to have ejected a jet of matter that shows up on long-exposure photographs. The galaxy is also a strong radio source. Another puzzling galaxy is M82 in Ursa Major, best found by tracking down its larger and brighter companion M81, which is a normal spiral galaxy of seventh magnitude. Both M81 and M82 can be glimpsed in binoculars under ideal conditions, and are easy enough in a 100 mm (4 inch) telescope. M82, which appears as an elongated smudge, has been described as an edge-on spiral encountering a cloud of dust in space, or as a peculiar galaxy undergoing an explosion. At present, astronomers do not fully understand what is happening in M82.

Although most galaxies are separated from one another by millions of light years, there are occasional 'traffic accidents'. This seems to be what is happening to M51 in Canes

Venatici, a beautiful face-on spiral galaxy popularly known as the Whirlpool. It is of historical interest as being the first galaxy in which spiral form was detected, but the spiral shape becomes evident only in apertures above about 250 mm (10 inches). Owners of small telescopes will be able to detect only the bright centre of the galaxy, but nearby will be visible the centre of another, smaller galaxy. Photographs show that this neighbour, NGC 5195, is actually distorting one of the spiral arms of M51 by its gravitational pull.

Some spiral galaxies have unusually active centres and are known as *Seyfert galaxies*. The brightest example, M77 in Cetus, is visible in a 100 mm (4 inch) telescope as a ninth-magnitude glow. Seyfert galaxies are related to the even more violently active objects known as *quasars*. Quasars remain virtually starlike in even the largest telescopes, and are now thought to be the brilliant centres of extremely remote galaxies. The brightest quasar, 3C 273 in Virgo, appears as a 13th-magnitude blue star. It lies about 3000 million light years away. The most distant quasars lie at the edge of the visible Universe, over 10,000 million light years away.

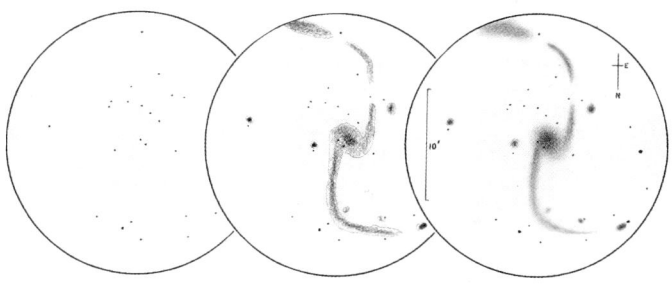

Deep-sky objects should be drawn in negative – that is, with the brightest parts darkest. Start by marking in the brightest stars in the field (a), and then with a soft pencil (4B) fill in the extent of the object (b). Smudge the pencil shading with your finger or with a soft, pencil-shaped tube called an artist's stub (this can be bought from a graphic arts supplier, or you can make your own from rolled-up blotting paper). Use a soft eraser to clean up the drawing and to emphasize certain areas such as dark lanes in the nebula or galaxy. Add more pencil shading as necessary to the brightest areas (c). Finish by making any notes about the object, as well as recording the usual details of date, time, instrument and magnification. A note on the field of view, as a guide to the size and orientation of the object, can also be useful. This is a drawing of the galaxy M33 in Triangulum.

ASTROPHOTOGRAPHY

You can photograph the night sky with most types of camera, the main exceptions being certain of the modern 'high-tech' automatic cameras. A single-lens reflex (SLR) camera with a mechanical shutter (rather than an electronic one) is by far the best for astrophotography, but interesting results can be obtained with even a simple 'point and click' camera. The crescent Moon in the evening twilight, perhaps with a bright planet such as Venus or Jupiter nearby, makes a pleasing view that can be captured with a simple camera. A series of photographs of two bright planets in the same part of the sky taken on successive nights will clearly show their movement relative to each other. You can photograph such subjects without using a telescope, but you can also try taking close-up views through a telescope or binoculars.

Pinkish prominences of hydrogen are visible at the limb of the eclipsed Sun on 1979 February 26, photographed through a pair of 16×50 binoculars with the set-up shown on pages 132/3.

To take a successful photograph of a celestial object, you will need to get as much light from it as possible onto the film, and this means a long exposure with a wide aperture. A simple camera may have controls that allow you to change the shutter speed or the lens aperture, or both. On some cameras this is done by moving a switch to different positions for different levels of ambient lighting or for different film speeds; you should choose the position suitable for cloudy conditions or the slowest speed. On some old types of simple camera with a built-in flash that uses bulbs, when a bulb is in place the camera shutter automatically changes to a longer exposure time suitable for flash photography. You can fool such a camera into giving you a longer exposure by inserting a flash bulb. Needless to say the flash itself will not enhance your photograph, so you should insert a spent bulb unless you deliberately want to illuminate the foreground.

You can try taking a magnified view of the Moon by holding the camera lens against the eyepiece of a telescope or one half of a pair of binoculars, having first focused the instrument on the Moon. Beware of touching the telescope, which may set up vibrations that will blur the picture. If the camera has an adjustable focus, ensure that it is focused on infinity, and if it has adjustable exposure time try 1/30 second or so. Even though the telescope may appear to be in focus to your eye, it may not be in focus for the camera, so try taking additional shots with the telescope focused either side of the best visual position. The resulting pictures should show the

A simple astrophotograph taken with basic equipment: the near-full Moon snapped with a fixed-focus compact camera through the eyepiece of a 100mm refractor on ISO 100 film.

Moon's major features such as its dark maria and bright rayed craters. As with all astrophotography, be prepared to experiment – and expect more failures than successes at first. Write down the various settings you use and see which works best.

Most forms of astrophotography require fast film, and in recent years some excellent fast films have become widely available, both colour and black-and-white. Film speeds are measured by an ISO rating (International Standards Organization). A film with double the ISO rating of another is twice as fast and hence requires half the exposure time to produce the same result. Typical film speeds for general use range from ISO 100 to ISO 400, but films with even faster speeds of ISO 1000 and above are available for both prints and transparencies, and are ideal for astrophotography.

Modern SLR cameras with autoexposure facility are not well suited to astrophotography, although some have a manual override that allows you to alter the settings. Many of these cameras will not allow exposures longer than about 1 second, and have zoom lenses that are limited to a maximum aperture of $f/4.5$. This restricts you to twilight shots of bright stars and planets or through-the-telescope snaps on very fast film.

The best type of camera for astrophotography is the older type of SLR camera on which the lens aperture and exposure time are freely adjustable. You can open the lens to its widest ($f/2$ say), and set the exposure time at, say, 1 second for a twilight view of the Moon and planets. Mount the camera

The crescent Moon and Venus in twilight make a simple but pleasing composition that can be snapped with a simple camera on fast film.

firmly on a tripod, or clamp or wedge it so that it cannot move when the shutter is pressed. A valuable accessory is a cable-release, which allows you to operate the shutter without touching the camera, thereby reducing the possibility of blurred pictures. If you want to take longer exposures you should use the B setting, so that the shutter will remain open for as long as you keep it pressed. A good tip when making exposures on the B setting is to cover the lens with a piece of card before opening the shutter. The card is moved away to start the exposure, and replaced over the lens before the shutter is closed. This procedure eliminates vibrations. Some SLR cameras allow you to lock the mirror, thus preventing mirror 'slam' which causes vibrations.

With these cameras it is possible to make extended time exposures of many seconds, or even minutes, using the B setting (note that long exposures drain the batteries of some cameras with electronic shutters). Mount the camera firmly, focus it on infinity and open the lens to full aperture. Point the camera at a constellation and leave the shutter open for about 20 seconds. The resulting picture will show stars fainter than those visible to the naked eye (exactly how faint will depend on the speed of the film). A series of such pictures taken on successive nights will help you to identify Uranus and Neptune and the brighter minor planets as they move against the background stars. Colour slide film is preferable to print film for such pictures since transparencies generally show sharper images and fainter stars than prints do. In

A motor-driven camera mount like the one shown on page 138 was used for this 6 minute exposure that shows the stars of Sagitta and Vulpecula, including the distinctive 'Coathanger' grouping. It was taken with a 55mm lens, aperture f/2.8, on ISO 1600 film.

addition, astrophotographs are often difficult to print successfully, and the colours on slide film are usually better. It is best to ask the processors not to mount the slides since they may accidentally cut the dark frames in half, unless you have some daylight shots at the start of the roll to guide them.

If you leave the shutter open for longer than about 20 seconds the star images will be drawn out into noticeable trails by the rotation of the Earth. An exposure of several

The stars of Orion, drawn out into trails as they rise, show various colours – notably the distinctive orange of Betelgeuse. The picture was taken with a 55mm lens, exposure 70 minutes on ISO 1600 film.

minutes will show clearly that stars near the celestial equator have fairly straight trails, whereas the trails of those near the pole are curved. The colours of various stars are much more pronounced on photographs than they are to the naked eye. Pictures of star trails around Polaris (as on pages 140/41) are always popular. You can add drama to your pictures by including a horizon with objects such as trees and hills.

If you live in a built-up area you will find that exposures of more than about five minutes become heavily fogged by streetlights. Special filters can be obtained to cut out artificial light. The best answer, though, is to find a darker site where you can take much longer exposures – particularly if a meteor shower is under way since you can observe visually while taking a series of long-exposure photographs in the hope of catching some of the brightest meteors on film. For such work, black-and-white film may be preferable to colour since it is cheaper and can be processed at home. It is wise to check the lens during an extended observing session since it may dew up and blur the photographs. A long lens hood will help reduce the build-up of dew.

One of the advantages of SLR cameras is that their lenses are interchangeable. The focal length of a standard lens is 50 mm, but you can change to a lens of shorter focal length for wider-angle views, or a telephoto lens of longer focal length. A wide-angle lens of, say, 35 mm focal length might be worth trying for meteors and aurorae, or for capturing the sweep of the Milky Way. With longer focal lengths the image scale is increased and the exposure time must be reduced

A motor-driven equatorial mount allows long exposures without star trailing.

accordingly to prevent star trailing. For example, a 300 mm lens is six times the focal length of a standard 50 mm lens, so the exposure time should be six times less to prevent star trailing, i.e. 3–4 seconds rather than 20 seconds.

Even more spectacular results can be obtained if the camera can be driven to follow the stars during the exposure. One way of doing this is to mount the camera piggyback on an equatorially mounted telescope, although motor-driven camera platforms can be bought. With such a set-up the star images can build up on the film over several minutes, allowing fainter objects to be recorded than with the star-trail method. This is a good way of photographing comets and the Milky Way, or hunting for novae.

Better-quality star images can be obtained by stopping down the lens slightly from its full aperture, say $f/2.8$ instead of $f/2$. Always make a note of the date and time of each exposure, since you never know what unexpected object you may have recorded on film.

A more advanced technique is to remove the lens from the camera entirely and fix the camera body to the eyepiece mount of the telescope, either with or without an eyepiece in place. Special adaptors for this purpose can be obtained from telescope or camera suppliers. Without the

Considerable detail is visible on the half-illuminated Moon, photographed through an 800mm telephoto lens at f/32 with a 1 second exposure on ISO 25 film.

eyepiece the field of view will be wider and the image brighter; this is known as *prime-focus photography* and is suitable for the whole disk of the Moon and for deep-sky objects. For close-ups of the Moon's surface or pictures of the planets you should photograph through an eyepiece, a technique known as *eyepiece projection*. The difficulty here lies in accurately focusing the image in the camera's viewfinder. You will also begin to encounter the effects of bad seeing at the higher magnifications obtainable with eyepiece projection – bad seeing will blur the image no matter how carefully you focus.

When photographing through a telescope you can try using a medium-speed or even a slow film, as these are less grainy than the faster films. Exposures of no more than a second or so will be needed for the Moon and planets with eyepiece projection (less in prime-focus photography). Always bracket your exposures – i.e. use a range of exposure times to ensure optimum results. An equatorially mounted telescope with a motor drive will be suitable to prevent image motion during brief exposures of the Moon and planets. However, for long exposures of objects such as faint comets, nebulae or galaxies some form of tracking control will be needed to ensure that the telescope remains accurately pointed at the target through the exposure, which may last several minutes or more.

You cannot carry out prime-focus or eyepiece-projection photography if your telescope lacks sufficient focusing range. So if you want to use a telescope for astrophotography, check first that its focusing range is suitable. Telescopes with open tubes have the advantage that it is often possible to move the secondary mirror closer to or away from the primary mirror to adjust the focus. Some telescopes allow you to move the mirror up and down the tube for the same purpose.

Star trails curve around the north celestial pole in this 66 minute exposure on ISO 1600 film with a 24 mm lens at f/2.8. Even Polaris has a short trail, since it is 1° from the exact north pole. Compare these trails with the nearly straight trails of the stars of Orion, on page 137, which lie close to the celestial equator.

Aratus Cilix

Taurus

Gemini

Caput medusae

Perseus

Cassiopeia

Erichthonius

Cancer

Vrsa maior

Vrsa minor

Leo

Draco

Bootes

Virgo

Seguus

M. Manli us Romanus

Libra

CONSTELLATIONS AND OBJECTS OF INTEREST

On the following pages are listed the constellations and the map(s) on which they are to be found. The main objects that are visible through binoculars or small telescopes ('small' meaning having an aperture of 50 to 75 mm/2 to 3 inches) are described, plus selected objects that require a larger instrument. In skies badly polluted by dirt and artificial lights, diffuse objects such as nebulae and galaxies will be difficult to see. When the atmosphere is unsteady, close double stars will be more difficult to split than in steady air. Stellar data in the notes are from the Yale Observatory *Bright Star Catalogue*, 4th edition. Data from other sources may differ, particularly for distances, which in most cases are not accurately known.

The maps on pages 166 to 175 show the whole sky in ten sections with stars down to fifth magnitude. The Milky Way is shown in light blue. Double and multiple stars are represented by special symbols, but not all of these stars have separations wide enough for them to be split in small telescopes. Deep-sky objects, also represented by special symbols, are labelled with their Messier (M) numbers where they have them, or by their NGC numbers (without prefix) or IC numbers (prefix I.). Constellation boundaries are marked as solid lines. Coordinates on the maps are right ascension (hours) and declination (degrees).

On pages 176 to 187 are twelve monthly maps depicting a band of sky stretching from north to south as it appears each month around 10 pm in mid-month (or 11 pm when daylight saving time is in operation). These charts will enable you to locate various constellations at a convenient time of night. For each two-hour time difference *after* 10 pm, go one map *forward*; for each two-hour time difference *before* 10 pm, go one map *back*. For example, to see how the sky appears at 6 pm in mid-December, go back two maps from the December map to the October one. The positions of the northern and southern horizons are shown as they appear from latitudes between the equator and 60°N; you can draw a pencil line across the maps to mark your own horizon. Since celestial objects are at their greatest altitude when due south, you will be able to see from these maps which stars and constellations are forever below your horizon. These monthly maps show fewer stars than the all-sky maps (down to fourth magnitude instead of fifth).

Andromeda

Maps: 8, 2, 3

γ **Andromedae** is a beautiful double star consisting of orange and blue components of mags. 2.3 and 4.8, divisible with small telescopes (see p. 107).

M31, the Andromeda spiral galaxy, is a twin of our own Galaxy and is over 2 million light years away. It is visible to the naked eye as an elongated blur. Binoculars show it clearly and give some idea of its true extent. In telescopes of 100 mm (4 inches) or more two elliptical satellite galaxies, **M32** and **NGC 205**, can be seen. M32 is the smaller and more easily visible of the pair. **NGC 7662** is a small but prominent planetary nebula, visible in small telescopes as a rounded, greenish disk of 9th mag. like an out-of-focus star.

Antlia

Map: 6

ζ **Antliae** is a pair of 6th-mag. stars wide enough to be split in binoculars. Small telescopes will split the brighter of the pair, ζ¹ Antliae, into two fainter stars.

Apus

Map: 9

Aquarius

Map: 3

ζ **Aquarii** is a close double consisting of near-identical white stars of 4th mag. that orbit each other every 850 years. They are too close to split in the smallest telescopes, but apertures of 75 mm (3 inches) should divide them with high

magnification (see p. 107).
M2 is a globular cluster visible as a fuzzy star in binoculars and small telescopes.
NGC 7009, the Saturn Nebula, is an 8th-mag. planetary nebula that gets its name from its resemblance to the planet Saturn when seen in apertures of 200 mm (8 inches) and above. In small telescopes it appears as an elliptical bluish disk.
NGC 7293, the Helix Nebula, is a large planetary nebula, almost half the apparent size of the Moon, but difficult to spot because its light is spread out. Binoculars and rich-field telescopes show it as a faint, rounded patch of light in clear, dark skies.

Aquila

Maps: 4, 3

α **Aquilae**, Altair, mag. 0.77, spectral class A7V, distance 16 light years.
η **Aquilae** is one of the brightest Cepheid variables, ranging from mag. 3.5 to 4.4 in a period of 7.18 days.

Ara

Map: 9

NGC 6193 is a scattered open cluster of binocular stars.
NGC 6397 is a large and loose globular cluster visible in binoculars.

Aries

Map: 8

γ **Arietis** is an attractive double consisting of twin white stars, each of mag. 4.8, that can be split in small telescopes.

Auriga

Maps: 7, 2

α **Aurigae**, Capella, mag. 0.08, distance 40 light years, is the sixth-brightest star in the sky. It is a spectroscopic binary consisting of two G-type giants that orbit each other in a period of 104 days, but do not eclipse.
ε **Aurigae** is a highly unusual eclipsing binary with the longest known period. Every 27 years it fades from mag. 2.9 to 3.8, where it remains for 14 months before returning to mag. 2.9. The main star is classified as FIa (a white supergiant) and its dark companion is believed to be a close binary star embedded in a disk of dust. The next eclipse is due to start in late 2009.
ζ **Aurigae** is another remarkable eclipsing binary, consisting of an orange giant star orbited by a smaller blue companion. When the larger star eclipses its companion, every 2 years 8 months, the combined brightness of ζ Aurigae drops from mag. 3.75 to 3.9 for six weeks, although this change is barely noticeable to the eye.
M36 is the middle of a row of

three impressive open clusters in Auriga, all of which appear comet-like in binoculars. **M36** is the smallest of the three, but is most easily resolved into stars by small telescopes. **M37** is the largest and densest of the three Auriga clusters, but it is the most difficult to resolve since its stars are faint.
M38 is a scattered open cluster of stars that looks rather like a starfish in small telescopes.

Boötes

Maps: 5, 1

α **Boötis**, Arcturus, mag. −0.04, spectral class K1III, distance 34 light years, is the fourth-brightest star in the sky.
ε **Boötis** is a beautiful but difficult double star consisting of a 3rd-mag. orange giant that overwhelms its 5th-mag. blue–white companion in small telescopes. An aperture of at least 75 mm (3 inches) and high magnification are needed to separate this unequal pair of contrasting colours (see p. 107).
ξ **Boötis** is an attractive pair of yellow and orange stars, mags. 4.7 and 6.9, that orbit each other every 150 years and can be split in small telescopes.

Caelum

Map: 7

Camelopardalis

Maps: 2, 1

Cancer

Map: 6

ζ **Cancri** is a double star with yellow components of mags. 5.1 and 6.2 that can be split in small apertures.
M44, the Beehive, is a large open cluster visible to the naked eye as a misty patch on dark nights and best seen in binoculars or rich-field telescopes on account of its large size – three times the apparent diameter of the Moon. It consists of a scattering of stars of 6th mag. and fainter. Alternative names are Praesepe and the Manger.
M67 is an open cluster, smaller and denser than the Beehive, visible in binoculars but needing a telescope to resolve its individual stars.

Canes Venatici

Maps: 5, 1

α **Canum Venaticorum**, Cor Caroli, is a double star of mags. 2.9 and 5.6, easily split in small telescopes.
M3 is a globular cluster visible as a fuzzy star in binoculars and small telescopes.
M51, the Whirlpool Galaxy, is a good example of a spiral galaxy viewed face-on. An irregular companion galaxy, **NGC 5195**, lies at the end of

one of its spiral arms. Small to medium telescopes show the nuclei of the two galaxies and a hint of their surroundings, but an aperture of at least 200 mm (8 inches) is needed to detect M51's spiral arms.

Canis Major

Map: 7

α **Canis Majoris**, Sirius, mag. −1.46, spectral class A1V, distance 8.6 light years, is the brightest star in the sky. It is orbited by a mag. 8.5 white dwarf, period 50 years. The white dwarf is difficult to see even in quite large telescopes because of the glare from Sirius itself.
M41 is a large open cluster well seen in binoculars and rich-field telescopes, and dimly visible to the naked eye under good conditions. Its stars appear to lie in zigzagging chains.
NGC 2362 is a small open cluster centred on the blue supergiant τ **Canis Majoris**, mag. 4.4, and is well seen in small telescopes.

Canis Minor

Map: 7

α **Canis Minoris**, Procyon, mag. 0.38, spectral class F5IV–V, distance 11.4 light years. Procyon has a white dwarf companion of mag. 10.3, orbital period 40 years, but only large telescopes will show it.

Capricornus

Map: 3

α **Capricorni** is a wide optical double, mags. 3.6 and 4.2, divisible in binoculars or even with sharp eyesight. The fainter star is a yellow supergiant, the brighter is an orange giant. Each star is itself a binary, but in both cases the companions are so faint that they are difficult to see in small telescopes.
β **Capricorni** is a wide double for small telescopes, with components of mags. 3.1 and 6.1.
M30 is a moderately bright globular cluster visible in small telescopes.

Carina

Map: 9

α **Carinae**, Canopus, mag. −0.72, spectral class F0II, distance 120 light years, is the second-brightest star in the sky, but is not visible from north of latitude 37°N.
η **Carinae** is a peculiar variable star. It is believed to be young, and so massive that it is unstable; it may become a supernova in the next 10,000 years. It is currently (1990) around 6th mag., but in 1843 it reached a peak of −0.8, outshining Canopus. It is a member of a cluster of stars within the nebula NGC 3372 (see p. 148).
NGC 2516 is a large cluster visible to the naked eye. Its brightest star is a red giant of mag. 5.2.

NGC 3114 is a large open cluster, good in binoculars, whose brightest members are of 6th mag.

NGC 3372 is a large bright nebula 2° across visible in binoculars around the star η Carinae. It includes a dark area near η Carinae called the Keyhole Nebula.

NGC 3532 is a bright, elliptical-shaped open cluster visible to the naked eye.

IC 2602 is a large and bright knot of naked-eye stars, rather like a mini-Pleiades, containing the mag. 2.8 star θ **Carinae**.

Cassiopeia

Map: 2

γ **Cassiopeiae** is an erratic variable of the type known as a *shell star* because it throws off gas from time to time. Its brightness has varied between mags. 1.6 and 3.0, and currently (1990) lies around mag. 2.5. Further changes could occur at any time.

η **Cassiopeiae** is an attractive double star for small telescopes, consisting of a mag. 3.5 yellow star with a wide mag. 7.5 orange companion.

ϱ **Cassiopeiae** is a semiregular G-type supergiant that varies between mags. 4.1 and 6.2 with no fixed period.

M52 is a rich cluster of faint stars, visible as a misty patch in binoculars but resolvable in small telescopes.

NGC 457 is a rectangular cluster that has been compared in shape to an owl, its two brightest stars marking the owl's eyes.

Centaurus

Maps: 9, 5, 6

α **Centauri**, Rigil Kentaurus, 4.3 light years away, is the closest naked-eye star to the Sun. At mag. −0.27 it is the third-brightest star in the sky, but small telescopes split it into a sparkling binary of mags. 0.0 and 1.3, spectral classes G2V and K1V. These two stars orbit each other every 80 years. A third member of the α Centauri system is the 11th-mag. red dwarf Proxima Centauri.

β **Centauri**, Hadar or Agena, mag. 0.61, spectral class B1III, distance 360 light years. With α Centauri it forms the 'pointers' to Crux, the Southern Cross.

ω **Centauri** is not a star but a globular cluster − the largest and brightest in the sky, appearing like a 4th-mag. star whose light is spread across a Moon's-breadth of sky. Its brightest stars can be resolved in small telescopes.

NGC 3918, the Blue Planetary, is an 8th-mag. planetary nebula similar in appearance to the planet Uranus but three times larger and hence easily detectable in small telescopes.

NGC 5128 is a remarkable elliptical galaxy, also known as the radio source Centaurus A. Small telescopes show it as a 7th-mag. rounded fuzz, while long-exposure photographs reveal that it is bisected by an unusual dark lane of dust, faintly visible in apertures above 100 mm (4 inches), that may have resulted from a merger with another galaxy.

Cepheus

Map: 2

β **Cephei**, mag. 3.2, has a wide 8th-mag. companion visible in small telescopes; both stars appear blue–white. β Cephei pulsates with a period of 4.5 hours, its brightness varying by 0.1 mag. – barely noticeable to the naked eye.
δ **Cephei** is the prototype Cepheid variable, ranging from mag. 3.5 to 4.4 with a period of 5.37 days. It is also an attractive double star for small telescopes. The variable star appears cream in colour, since it is a yellow supergiant (spectral class ranging from F5Ib to G2Ib as it pulsates), while the companion is mag. 6.3 and bluish. For a comparison chart, see p. 112.
μ **Cephei**, the Garnet Star, is a red supergiant whose strong colour is clearly noticeable in binoculars. It is a semiregular variable, ranging between mags. 3.4 and 5.1 with a period of about two years.

Cetus

Map: 8

o **Ceti**, Mira, is a red giant, the prototype of the common long-period variables. Mira itself has a period of about 11 months during which it varies from as bright as 2nd mag. to as faint as 10th mag. For a comparison chart, see p. 111.
τ **Ceti**, mag. 3.5, spectral class G8V, distance 11.4 light years, is one of the most Sun-like of the nearby stars.
M77 is the brightest of the Seyfert galaxies, spirals with unusually bright centres that are related to quasars. M77 itself appears as a 9th-mag. fuzzy spot, though only its bright centre is well seen in small telescopes.

Chamaeleon

Map: 9

Circinus

Map: 9

Columba

Map: 7

Coma Berenices

Map: 5

The **Coma star cluster** consists of a triangular scattering of 5th- and

6th-mag. stars covering 10°.

24 Comae Berenices is a wide and attractive double star, easy in small telescopes, with yellow and white components of mags. 5.0 and 6.6.

M64, the Black Eye Galaxy, is the brightest of numerous galaxies in the constellation. It is a spiral galaxy of mag. 8.5. A dark dust patch near the nucleus produces the 'black eye' effect, although this will probably not be visible in apertures below 100 mm (4 inches), depending on sky conditions.

NGC 4565 is a famous example of a spiral galaxy seen edge-on. Apertures of 100 mm (4 inches) show it as a faint, elongated smudge, while in larger telescopes a lane of dust can be seen running along it.

Corona Australis

Map: 4

NGC 6541 is a 7th-mag. globular cluster visible in binoculars and small telescopes.

Corona Borealis

Map: 5

α **Coronae Borealis**, Gemma, is an eclipsing binary star of the Algol type. Every 17.4 days it goes from mag. 2.2 to 2.3, and back again, although this change is barely perceptible to the eye.

R Coronae Borealis is the prototype of an unusual class of variable stars that undergo sudden drops in brightness, apparently due to the build-up and subsequent dispersal of a dust shell. R Coronae Borealis is a G-type supergiant that is normally of 6th mag., but can unexpectedly plummet within weeks to as faint as 15th mag., remaining there for months before brightening again, often irregularly.

T Coronae Borealis, the Blaze Star, is a recurrent nova which in 1866 and 1946 flared up to 2nd mag. Usually it is around 10th mag.

Corvus

Map: 5

δ **Corvi** is a blue–white star of mag. 3.0 with a wide mag. 8.3 companion detectable in small telescopes.

Crater

Map: 6

Crux

Map: 9

α **Crucis**, Acrux, mag. 0.79, distance 400 light years, is split by small telescopes into a pair of sparkling blue–white stars, mags. 1.3 and 1.7.

NGC 4755, the Jewel Box, is a colourful open cluster also known as the κ Crucis cluster. Its brightest stars are B-type supergiants of 6th mag.

The **Coalsack** is a celebrated

dark nebula blanketing an area of Milky Way 7° long and 5° wide.

Cygnus

Maps: 2, 3, 4

α **Cygni**, Deneb, mag. 1.25, spectral class A2Ia, is the most distant of the first-magnitude stars, about 2000 light years from us (the exact value is uncertain). It is one of the most luminous stars known, with an estimated brilliancy of 60,000 Suns.
β **Cygni**, Albireo, is a fine example of a coloured double star, divisible with small telescopes (or even good binoculars) into amber and blue–green components of mags. 3.1 and 5.1 (see p. 107).
o¹ **Cygni** is a beautiful binocular pair of yellow and blue stars, mags. 3.8 and 4.8, like a wide version of Albireo. Good binoculars or small telescopes show another blue star, mag. 7.0, much closer to the yellow one.
χ **Cygni** is a variable red giant like Mira that ranges between mags. 3 and 14 with a period of 400 days.
61 Cygni is an attractive pair of orange stars, mags. 5.2 and 6.0, easily split in small telescopes (see p. 107). They orbit each other with a period of 650 years.
M39 is a bright but loose open cluster, best seen in binoculars and rich-field telescopes.
NGC 6826, the 'blinking

planetary', is a planetary nebula visible in 75 mm (3 inch) telescopes, so named because to some observers it appears to blink on and off when they look alternately towards and away from it.
NGC 7000, the North America Nebula, is a bright nebula shaped like the continent after which it is named. It shows up well on long-exposure photographs but is not easy to see visually. Binoculars or a rich-field telescope on a dark night are needed to spot it, but a nebula filter will make it stand out nicely.
The **Veil Nebula** is the remains of a supernova that exploded perhaps 60,000 years ago. The wreckage forms twisted veils of gas the brightest of which, **NGC 6992**, can just be detected in binoculars under ideal conditions, and is best seen with a nebula filter.

Delphinus

Map: 3

γ **Delphini** is an attractive binary for small telescopes, consisting of yellow and white stars of mags. 4.3 and 5.1. Adding to the attraction in the same telescopic field of view is the closer and fainter double designated Struve 2725, mags. 7.6 and 8.4.

Dorado

Maps: 9, 10

β **Doradus** is a bright Cepheid variable that pulsates with a period of 9.84 days between mags. 3.5 and 4.1.

The **Large Magellanic Cloud** (LMC) is the larger and nearer of the two companion galaxies to our Milky Way; it lies 160,000 light years away and has an estimated mass of 10 billion Suns. The LMC is usually classified as an irregular galaxy, but it does have a trace of barred-spiral structure. As a mini-galaxy it contains numerous star clusters and nebulae, the brightest of which is the spidery-shaped Tarantula Nebula, **NGC 2070**, visible as a fuzzy star to the naked eye.

Draco

Maps: 1, 2

μ **Draconis** is a tight double star that can be split in a 100mm (4 inch) telescope with high magnification. It consists of two identical stars of mag. 5.7 orbiting each other in 480 years.

ν **Draconis** is a wide pair of identical mag. 4.9 stars neatly divided in binoculars and easy in the smallest of telescopes.

ψ **Draconis** is a wide double that a small telescope easily splits into components of mags. 4.6 and 5.8.

16 and **17 Draconis** appear in binoculars as a wide pair of mags. 5.5 and 5.1. Small telescopes with high magnification show a mag. 6.5 star close to the brighter of the pair.

39 Draconis appears of mag. 5.0 to the naked eye; binoculars or small telescopes bring into view a wide companion of mag. 7.1. Higher magnification reveals that the brighter star is in fact a tight double, making this a triple system.

NGC 6543, a relatively bright planetary nebula, shows a blue–green disk of 9th mag. in small telescopes under high magnification.

Equuleus

Map: 3

Eridanus

Maps: 7, 8, 10

α **Eridani**, Achernar, mag. 0.46, spectral class B3V, distance 125 light years.

θ **Eridani** is a bright and impressive pair of blue–white stars, mags. 3.2 and 4.4, neatly split in small telescopes.

o^2 **Eridani** (also designated 40 Eridani) is a triple star containing the most easily visible white dwarf in the sky, and a red dwarf. A small telescope will just show the mag. 9.5 white dwarf some way from the mag. 4.4 yellow primary. Larger apertures are needed to show the 11th-mag. red dwarf close to the white dwarf.

32 Eridani is a beautiful

coloured double star for small telescopes, consisting of an orange giant of mag. 4.8 with a mag. 6.1 companion that appears blue–green by contrast.

Fornax

Map: 8

Gemini

Map: 7

α **Geminorum**, Castor, appears to the naked eye as a blue–white star of mag. 1.6, but small to medium telescopes with high magnification split it into a tight double of mags. 1.9 and 2.9 (see p. 107). These two stars orbit around each other once every 500 years. There is also a much wider 9th-mag. companion. Remarkably, each of these three visible stars is itself a spectroscopic binary, making Castor a six-star family. Castor lies 50 light years away.
β **Geminorum**, Pollux, mag. 1.14, spectral class K0III, distance 35 light years.
η **Geminorum** is a red giant semiregular variable that fluctuates between mags. 3.3 and 3.9 with a period of 8 months.
M35 is a large, bright open cluster the apparent size of the Moon, just resolved in binoculars and impressive in small telescopes.

Grus

Maps: 1, 3

δ **Gruis** is an optical double, a pair of unrelated yellow stars, mags. 4.0 and 4.1, individually distinguishable to the naked eye.
μ **Gruis** is another optical double, mags. 4.8 and 5.1.

Hercules

Maps: 4, 1, 2

α **Herculis** is a red supergiant semiregular variable with a range from mag. 2.7 to 4.0; small telescopes show a yellow companion of mag. 5.4.
ζ **Herculis** is a close double with an orbital period of 34.5 years (see the diagram on p. 109). To the naked eye it appears as a yellow–white star of mag. 2.8. Apertures of 100 mm (4 inches) and above split it into components of mags. 2.9 and 5.5.
ρ **Herculis** is a neat pair of blue–white stars, mags. 4.5 and 5.5, for small telescopes.
95 Herculis is an easy and attractive double for small telescopes, mags. 5.0 and 5.2, colours blue–white and orange.
M13 is the best globular cluster in northern skies, visible to the naked eye as a hazy star on clear nights and easily spotted in binoculars. Small to medium apertures show its brightest stars arranged in long chains.
M92 is another globular

cluster, somewhat smaller and fainter than M13. It can be seen in binoculars.

Horologium

Map: 10

Hydra

Maps: 6, 5

ε **Hydrae** is a close double, difficult to split in apertures smaller than about 75 mm (3 inches) because of the considerable difference in the stars' magnitudes, 3.4 and 6.8.

R Hydrae is a Mira-type red giant variable that ranges between 4th and 10th mags. over a period of 13 months; this period is gradually shortening.

U Hydrae is a semiregular variable giant star, deep red in colour, that ranges between 4th and 6th mags. with a period of about 15 months.

M48 is an open cluster about the same apparent size as the Moon, well seen in small telescopes at low power and easily detectable in binoculars.

M83 is an 8th-mag. spiral galaxy presented face-on to us. In small to medium apertures it appears elongated because of a central bar of stars.

NGC 3242, known as the Ghost of Jupiter, is a 9th-mag. planetary nebula that appears in small

telescopes as a blue–green disk of similar apparent size to Jupiter, hence its name.

Hydrus

Map: 10

Indus

Map: 10

Lacerta

Maps: 2, 3

Leo

Map: 6

α **Leonis**, Regulus, mag. 1.35, spectral class B7V, distance 70 light years.

γ **Leonis**, of mag. 1.9, has an unrelated star nearby, **40 Leonis**, mag. 4.8, divisible with good eyesight or in binoculars. Small telescopes split γ Leonis into a handsome pair of golden-yellow stars, mags. 2.2 and 3.5.

ι **Leonis** is a tight binary with a period of 192 years, unsuitable for small telescopes because of the closeness of the stars and their difference in magnitudes, 4.0 and 6.7. This is a good challenge for a 100 mm (4 inch) instrument, becoming somewhat easier with time as the two stars are gradually moving apart.

R Leonis is a long-period pulsating red giant like Mira.

It has a period of 10 months and ranges from 5th to 10th mag.

M65 and **M66** are a pair of 9th-mag. spiral galaxies, seen at an angle so that they appear as elongated smudges. They should be visible in small telescopes or even, under good conditions, large binoculars.

M95 and **M96** are another pair of spiral galaxies, fainter and more difficult than M65 and M66.

Leo Minor

Map: 6

Lepus

Map: 7

γ **Leporis** is a double, with components of mags. 3.6 and 6.2, that can be split in binoculars.

R Leporis is a Mira-type red giant variable with a period of 14 months or so. At maximum it can reach mag. 5.6, but at minimum it is as faint as 12th mag. It is known as Hind's Crimson Star on account of its deep red colour.

M79 is an 8th-mag. globular cluster visible as a fuzzy star in small telescopes. On the edge of the same low-power field lies the triple star h3752 consisting of a close pair of mags. 5.4 and 6.6 and a wider companion of mag. 9.1.

Libra

Map: 5

α **Librae** is a blue–white star of mag. 2.8; nearby is an unrelated star of mag. 5.2 that is visible to the naked eye or in binoculars.

δ **Librae** is an eclipsing binary that varies from mag. 4.9 to 5.9 in a period of 2.33 days.

ι **Librae** appears as a wide double of mags. 4.5 and 6.1 in binoculars. Telescopes show that the brighter star has a wide companion of 10th mag., which high magnification reveals is itself a close binary.

μ **Librae** is a close double of mags. 5.8 and 6.7, requiring a 75 mm (3 inch) aperture to split it.

Lupus

Maps: 9, 5

κ **Lupi** is a wide and easy double star for small telescopes, mags. 3.9 and 5.7.

μ **Lupi** appears in small telescopes as a mag. 4.3 star with a wide companion of mag. 7.2. High magnification on telescopes of 100 mm (4 inches) and above will show that the brighter of the two is itself a close double, consisting of near-identical stars of mags. 5.1 and 5.2.

ξ **Lupi** is an easy binary for small telescopes, consisting of components of mags. 5.1 and 5.6.

π **Lupi** is a close binary consisting of near-identical

stars of mags. 4.6 and 4.7, requiring an aperture of 75 mm (3 inches) to split.

Lynx

Map: 1

12 Lyncis is a close double of mags. 5.4 and 6.0, divisible with apertures of 75 mm (3 inches) with a wider mag. 7.1 companion.
19 Lyncis is an easy wide double of mags. 5.6 and 6.5, divisible in the smallest apertures.
38 Lyncis is a challenging double for small apertures, consisting of closely spaced components of mags. 3.9 and 6.6 that a 75 mm (3 inch) telescope should split.

Lyra

Map: 4

α **Lyrae**, Vega, mag. 0.03, spectral class A0V, distance 25 light years, is the fifth-brightest star in the sky.
β **Lyrae** is both a double and a variable star. Small telescopes show a wide companion of 9th mag. The brighter star, which appears cream in colour, is an eclipsing binary that varies between mags. 3.3 and 4.3 over a period of 12.9 days.
δ **Lyrae** consists of two unrelated stars of mags. 4.3 and 5.6, separately visible in binoculars. The brighter star is a red giant that is slightly variable.
ε **Lyrae** appears in binoculars like a glinting pair of cat's eyes, both of 5th mag. (see p. 107). Telescopes with high magnification reveal that each star is itself a close double, making this a rare and striking example of a quadruple star, hence its popular name, the 'double double'. The stars are difficult to split in the smallest telescopes because of their closeness.
ζ **Lyrae** is a binocular pairing of mags. 4.4 and 5.7.
M57, the Ring Nebula, is a well-known planetary nebula conveniently situated midway between β and γ Lyrae. Small telescopes show it as an elliptical 9th-mag. disk, and in larger telescopes the famous 'smoke ring' shape becomes more apparent.

Mensa

Map: 9

Microscopium

Map: 3

Monoceros

Map: 7

β **Monocerotis** is an outstanding triple star, separable by small telescopes into components of mags. 4.6, 5.4 and 5.6, the two fainter stars being the closest together.
ε **Monocerotis** is an easy pair for small telescopes, of

mags. 4.4 and 6.7.

M50 is a rich and bright open cluster, visible in binoculars and resolvable in small apertures.

NGC 2232 is a scattered open cluster whose brightest star is **10 Monocerotis**, mag. 5.1. The cluster is visible in binoculars.

NGC 2244 is a rectangular-shaped open cluster of binocular stars at the heart of the large, faint Rosette Nebula (**NGC 2237/39**). Under clear, dark skies the Nebula can be seen in binoculars as a pale disk more than a Moon's-breadth across.

NGC 2264 is a scattered open cluster appearing somewhat like a Christmas tree in small telescopes. Its brightest star is the blue–white **S Monocerotis**, mag. 4.7, which has a close companion of mag. 7.6. The cluster is surrounded by faint nebulosity, which includes the Cone Nebula, and shows up well only on long-exposure photographs.

Musca

Map: 9

Norma

Map: 9

$\gamma^{1,2}$ **Normae** is a naked-eye double – two unrelated yellow stars of mags. 4.0 and 5.0. ε **Normae** is a wide double for the smallest apertures, mags. 4.5 and 7.5.

ι^1 **Normae**, mag. 4.6, has a mag. 7.5 companion visible in small telescopes.

NGC 6087 is an open cluster visible in binoculars.

Octans

Map: 9

σ **Octantis** is the nearest naked-eye star to the south celestial pole. It is of mag. 5.5, and lies just over 1° from the pole.

Ophiuchus

Map: 4

ϱ **Ophiuchi** is an excellent multiple star for small telescopes. A low magnification shows it as a mag. 4.6 star flanked by wide companions of mags. 6.6 and 7.1, and a high magnification splits the main star into two components of mags. 5.0 and 5.9.

36 Ophiuchi is an attractive double star consisting of matching orange dwarf stars, each of mag. 5.1.

70 Ophiuchi is a well-known binary consisting of golden-yellow and orange dwarfs of mags. 4.2 and 6.0 with an orbital period of 88 years (see the diagram on p. 109). The components are now widening rapidly and by the year 2000 should be divisible in a 60 mm (2.4 inch) telescope with sufficient magnification.

M10 is a 7th-mag. globular cluster visible in binoculars.

M12 is another 7th-mag.

globular cluster visible in binoculars.

NGC 6633 is a scattered open cluster for binoculars.

IC 4665 is a large and loose open cluster well seen in binoculars.

Orion

Map: 7

α **Orionis**, Betelgeuse, is a red supergiant semiregular variable that ranges between about mags. 0.4 and 1.3 in a period of 6.4 years or so, with shorter variations of about a year superimposed. It lies about 650 light years away.

β **Orionis**, Rigel, mag. 0.12, spectral class B8Ia, is the seventh-brightest star in the sky. Its distance is not well determined, but is about 1000 light years.

θ^1 **Orionis**, the Trapezium, is a bright multiple star at the heart of the Orion Nebula. Small telescopes show four stars here, of mags. 5.1, 6.7 6.7 and 8.0. The light from these stars makes the Orion Nebula shine.

θ^2 **Orionis**, a close neighbour of θ^1 Orionis, is a binocular double of mags. 5.1 and 6.4.

ι **Orionis** is a double star for small telescopes, mags. 2.8 and 7.4, on the southern edge of the Orion Nebula. In the same field under low magnification can be seen the wide and easy double star designated Struve 747, mags. 4.8 and 5.7.

· σ **Orionis** is a notable multiple star. Small telescopes

show it as quadruple, with components of mags. 3.8, 8.8, 6.6 and 6.7 (in order of increasing distance from the brightest star). As a bonus, in the same field of view lies a faint triple star, Struve 761.

M42, the Orion Nebula, is the finest nebula in the sky – a huge cloud of glowing gas about 1400 light years away, easily visible to the naked eye and prominent in binoculars and small telescopes. A smaller patch of nebulosity, **M43**, adjoins it to the north. The whole complex looks like a swirling greenish fog covering twice the apparent diameter of the Moon, and is a wondrous sight in any aperture.

NGC 1977 is a bright nebula directly above the Orion Nebula, containing two 5th-mag. stars, and visible in binoculars. Two detached portions of the nebulosity bear the separate numbers **NGC 1973** and **NGC 1975**.

NGC 1981 is a large and loose cluster of binocular stars north of the Orion Nebula.

Pavo

Maps: 10, 9

$\varkappa$ **Pavonis** is a Cepheid variable that ranges between mags. 3.9 and 4.8 over a 9.1 day cycle.

NGC 6752 is a 5th-mag. globular cluster visible in binoculars.

Pegasus

Map: 3

β **Pegasi** is a red giant irregular variable that ranges between mags. 2.3 and 2.7. **M15** is a 6th-mag. globular cluster visible in binoculars as an enlarged star, less fuzzy than most globulars.

Perseus

Maps: 2, 7

β **Persei**, Algol, is the prototype eclipsing binary star. It varies between mags. 2.1 and 3.4 with a period of 2.87 days. For a comparison chart, see p. 114.
ϱ **Persei** is a red giant semiregular variable that ranges between mags. 3.3 and 4.0. Its period is about 50 days.
M34 is a large open cluster easily seen in binoculars and resolved into individual stars in small telescopes.
NGC 869 and **NGC 884**, the Double Cluster, are a pair of bright open clusters lying among the star fields of the Milky Way and visible to the naked eye. Binoculars or a rich-field telescope will show both of them in the same field of view. NGC 869 is the brighter and richer of the pair but NGC 884 contains some red giants among the otherwise blue–white stars.

Phoenix

Maps: 10, 3

β **Phoenicis** is a close double star of mags. 4.0 and 4.2; an aperture of at least 75 mm (3 inches) is needed to split it.
ζ **Phoenicis** is an Algol-type eclipsing binary that varies between mags. 3.9 and 4.4 in a period of 1.67 days. It has a 7th-mag. companion visible in small telescopes.

Pictor

Map: 9

Pisces

Maps: 8, 3

α **Piscium** is a challenging binary, mags. 4.2 and 5.2, requiring an aperture of at least 75 mm (3 inches) to split it. The orbital period is nearly 1000 years.
ζ **Piscium** is a wide and easy double for small telescopes, mags. 5.2 and 6.3.
ψ^1 **Piscium** is another wide and easy double, mags. 5.3 and 5.6, for the smallest apertures – or even powerful binoculars.
TX Piscium is a red giant irregular variable that ranges between mags. 4.8 and 5.2 and is distinguished by its deep red colour.

Piscis Austrinus

Map: 3

α **Piscis Austrini**, Fomalhaut, mag. 1.16, spectral class A3V, distance 22 light years.

β **Piscis Austrini** is a wide double for small telescopes with a pronounced magnitude difference, 4.4 and 7.9.

Puppis

Maps: 7, 9

k Puppis is a near-identical duo of blue–white stars, mags. 4.5 and 4.6, comfortably split in small apertures.

L^2 **Puppis** is a red giant semiregular variable, ranging between mags. 2.6 and 6.2 with a period of about 140 days. It forms a naked-eye double with L^1 **Puppis**, an unrelated star of mag. 4.9.

V Puppis is an eclipsing binary that ranges between mags. 4.4 and 4.9 with a 1.45 day period.

M46 is an open star cluster that appears as a misty patch in binoculars.

M47 is a scattered open cluster of bright stars, visible to the naked eye as a bright knot in the Milky Way, and totally different in character from nearby M46.

M93 is an open cluster visible in binoculars, triangular in shape.

NGC 2477 is a large and impressive open cluster which in binoculars looks like a globular cluster.

NGC 2451 is a large naked-eye cluster whose brightest star is a mag. 3.6 orange supergiant, **c Puppis**.

Pyxis

Map: 6

T Pyxidis is a recurrent nova that flared up to 6th mag. in 1890, 1902, 1920, 1944 and 1966, and is likely to undergo further outbursts.

Reticulum

Map: 10

Sagitta

Map: 4

M71 is a small 8th-mag. globular cluster for small telescopes.

Sagittarius

Map: 4

β **Sagittarii** is a naked-eye double of unrelated stars, mags. 4.0 and 4.3. The brighter of the pair, $β^1$ Sagittarii, has a 7th-mag. companion visible in small telescopes.

M8, the Lagoon Nebula, is an outstanding nebula bright enough to been seen with the naked eye and prominent in binoculars. It is elongated in shape and has a backbone of stars that is visible through binoculars.

M17, the Omega Nebula, is a

large wedge-shaped patch of nebulosity visible in binoculars.

M20, the Trifid Nebula, is an 8th-mag. nebula just detectable in binoculars. Telescopes show the dark lanes that trisect the nebula and give rise to its name. At its centre lies a 7th-mag. double star.

M22 is a superb 5th-mag. globular cluster, large and prominent in binoculars.

M23 is an elongated open cluster visible in binoculars.

M24 is a beautiful large star cloud in the Milky Way, appearing grainy in binoculars.

M25 is a large and scattered open cluster, well shown by binoculars. It contains the Cepheid variable **U Sagittarii**, which varies between mags. 6.3 and 7.1 with a period of 6.74 days.

Scorpius

Map: 4

α **Scorpii**, Antares, is a red supergiant semiregular variable that ranges between mags. 0.9 and 1.8 with a period of nearly 5 years.

β **Scorpii** is an easy optical double for small telescopes. The components, mags. 2.6 and 4.9, are both blue–white.

ν **Scorpii** may be split in small telescopes – or even binoculars – into a wide and easy double, mags. 4.0 and 6.3. A 75 mm (3 inch) telescope will show that the fainter component is itself double, mags. 6.9 and 7.9. The brighter component is also double, though much tighter, requiring an aperture of at least 150 mm (6 inches) to split, completing a quadruple grouping.

ξ **Scorpii**, mag. 4.2, has a mag. 7.3 companion visible in small telescopes. In the same field of view is a faint but easy double, mags. 7.4 and 8.1, which is related to ξ Scorpii, making this another quadruple star.

M4 is a 6th-mag. globular cluster, visible in binoculars but less easy than its magnitude might suggest because it is large and diffuse.

M6 is an open cluster just resolvable in binoculars and excellent in small telescopes. The brightest star is the orange supergiant semiregular variable **BM Scorpii**, which ranges between mags. 5.0 and 6.9 in about 2.3 years.

M7 is a spectacular open cluster, twice the apparent diameter of the Moon and visible to the naked eye. Binoculars easily resolve it into stars, arranged in chains against a starry Milky Way background.

NGC 6231 is a bright and compact open cluster in a rich area of the Milky Way. It is visible in binoculars but is best seen in a small telescope, in which it resembles a mini-Pleiades. Its brightest star is of mag. 5.5.

Sculptor

Maps: 8, 3

NGC 253 is a 7th-mag. spiral galaxy seen at quite a low angle to its plane, so that it appears cigar-shaped. It is visible in binoculars, and telescopes show signs of dark dust clouds in the spiral arms.

Scutum

Map: 4

M11, the Wild Duck Cluster, is a 6th-mag. open cluster visible as a smudgy patch in binoculars. But its real glory is apparent only in telescopes, which shows its chevron shape, like a flight of ducks, with a brighter orange star at the apex.

Serpens

Maps: 4, 5

δ **Serpentis** is a tight binary, mags. 4.2 and 5.2, that provides a challenge for the smallest apertures.
θ **Serpentis** is a pair of white stars, mags. 4.6 and 5.0, comfortably divisible in the smallest telescopes.
M5 is a 6th-mag. globular cluster visible as a fuzzy star in binoculars and small telescopes.
M16 is a compact open cluster visible in binoculars and small telescopes. It appears somewhat hazy but it lies within the faint Eagle Nebula, which shows up well only on long-exposure photographs.

Sextans

Map: 6

Taurus

Maps: 7, 8

α **Tauri**, Aldebaran, mag. 0.85, spectral class K5III, distance 60 light years. It is slightly variable.
θ **Tauri** is a wide double of mags. 3.4 and 3.8, visible as two stars with the naked eye.
λ **Tauri** is an Algol-type eclipsing binary that ranges between mags. 3.4 and 3.9 with a period of 3.95 days.
M1, the Crab Nebula, is one of the most celebrated objects in the sky, but is not easy to see in small instruments since it is only of 8th mag. Small telescopes will show it as a pale, elliptical nebulosity halfway in apparent size between Jupiter and the Moon.
M45, the Pleiades, is a famous open cluster of which about seven stars can be seen with normal eyesight, hence its popular name the Seven Sisters. Dozens more members are visible through binoculars. The brightest member of the cluster is η **Tauri**, Alcyone, a blue giant of mag. 2.9. The whole cluster is enveloped in faint nebulosity that shows up on long-exposure photographs. The **Hyades** is a very large, scattered cluster of naked-eye stars with no Messier or NGC designation. The stars are arranged in a V-shape and include the wide double θ Tauri.

Telescopium

Map: 10

δ **Telescopii** is a wide pair of unrelated blue–white stars, mags. 5.0 and 5.1.

Triangulum

Map: 8

ι **Trianguli** (also designated 6 Trianguli) consists of an orange giant of mag. 5.3 with a close bluish companion of mag. 6.9 visible in small telescopes.

M33 is a neighbouring spiral galaxy 2.6 million light years away, slightly farther away than the famous Andromeda galaxy but far more difficult to see because of its low surface brightness. Binoculars and rich-field telescopes show it as a large, pale, elliptical smudge.

Triangulum Australe

Map: 9

NGC 6025 is an open cluster visible in binoculars and small telescopes.

Tucana

Map: 10

β **Tucanae** is a naked-eye double of mags. 3.7 and 5.1. Small telescopes show that the brighter component is itself double, mags. 4.4 and 4.5.

κ **Tucanae** is a double star neatly split by small telescopes into components of mags. 5.1 and 7.3. Nearby in the same field lies a related 7th-mag. star.

47 Tucanae is the second-finest globular cluster in the sky, visible to the naked eye as a fuzzy star of 4th mag. At 15,000 light years away, it is one of the closest globulars to us.

NGC 362 is a 7th-mag. globular cluster visible in binoculars, lying near the Small Magellanic Cloud but actually part of our own Galaxy.

The **Small Magellanic Cloud** (SMC) is the smaller and more distant of the two satellite galaxies of our Milky Way, and lies about 190,000 light years away. The SMC appears to the naked eye as an elongated misty patch like a detached portion of the Milky Way. Binoculars and small telescopes show considerable detail in the SMC.

Ursa Major

Maps: 1, 6

ζ **Ursae Majoris**, Mizar, a blue–white star of mag. 2.3, is one of the best-known multiple stars in the sky. Keen eyesight reveals a wide companion, **80 Ursae Majoris**, Alcor, of mag. 4.0. A small telescope will show that Mizar has a closer companion, also of mag. 4.0 (see p. 107). All three stars are related and all are spectroscopic binaries.

ξ **Ursae Majoris** is a binary
with a period of 60 years,
mags. 4.3 and 4.8, colours
yellow and blue. By the year
2000 the two stars will be
divisible in apertures of
75 mm (3 inches) and
thereafter will become
progressively easier to split
(see the diagram on p. 109).
M81 is a spiral galaxy, seen at
an angle, that appears as a
7th-mag. elliptical patch in
small telescopes. Half a
degree to the north lies **M82**,
a smaller and fainter irregular
galaxy, elongated in shape.
Both are visible in binoculars
under good conditions, but
are not impressive.
M101 is an 8th-mag. spiral
galaxy seen face-on, visible as
a large but pale disk in
binoculars and small
telescopes under good
conditions.

Ursa Minor

Maps: 1, 2

α **Ursae Minoris**, Polaris, the
north pole star, is a yellow
supergiant Cepheid that varies
between mags. 1.9 and 2.1
with a period of 4 days.
Polaris lies about 500 light
years away. It has an
8th-mag. companion that is
visible in small telescopes.

Vela

Map: 9

γ **Velorum** can be split by the
smallest telescopes – or even
good binoculars – into
blue–white components of
mags. 1.8 and 4.3. Small
telescopes show a wider,
related companion of
mag. 7.7. A still wider
companion of mag. 9.1 is
unrelated.
NGC 2547 is an open cluster
for binoculars and small
telescopes.
IC 2391 is a large, scattered
open cluster, visible to the
naked eye, and ideal for
binoculars. Its brightest star is
of mag. 3.6

Virgo

Map: 5

α **Virginis**, Spica, mag. 0.98,
is a blue–white star 140 light
years away.
γ **Virginis** is a binary with a
171-year period, and consists
of two identical white stars of
mag. 3.5. As the diagram on
p. 109 shows, the two stars
are presently drawing together
and are becoming
increasingly
difficult to split.
M87 is a 9th-mag. elliptical
galaxy and is the easiest
member of the **Virgo Cluster**
to see. Other prominent
members of the cluster are
M49 and **M60**. The Virgo
Cluster lies 65 million light
years away.
M104, the Sombrero Galaxy,
is an 8th-mag. spiral galaxy

seen almost edge-on. In apertures over 150 mm (6 inches) it appears like a sombrero hat, but smaller apertures show it only as an elliptical smudge. It is not a member of the Virgo Cluster, but lies closer to us.

Volans

Map: 9

Vulpecula

Maps: 3, 4

M27, the Dumbbell Nebula, is reputedly the most conspicuous of all planetary nebulae in small instruments. It appears as a rounded, misty patch in binoculars. Small telescopes show an elliptical shape, while larger apertures show the double-lobed structure that gives it its popular name.

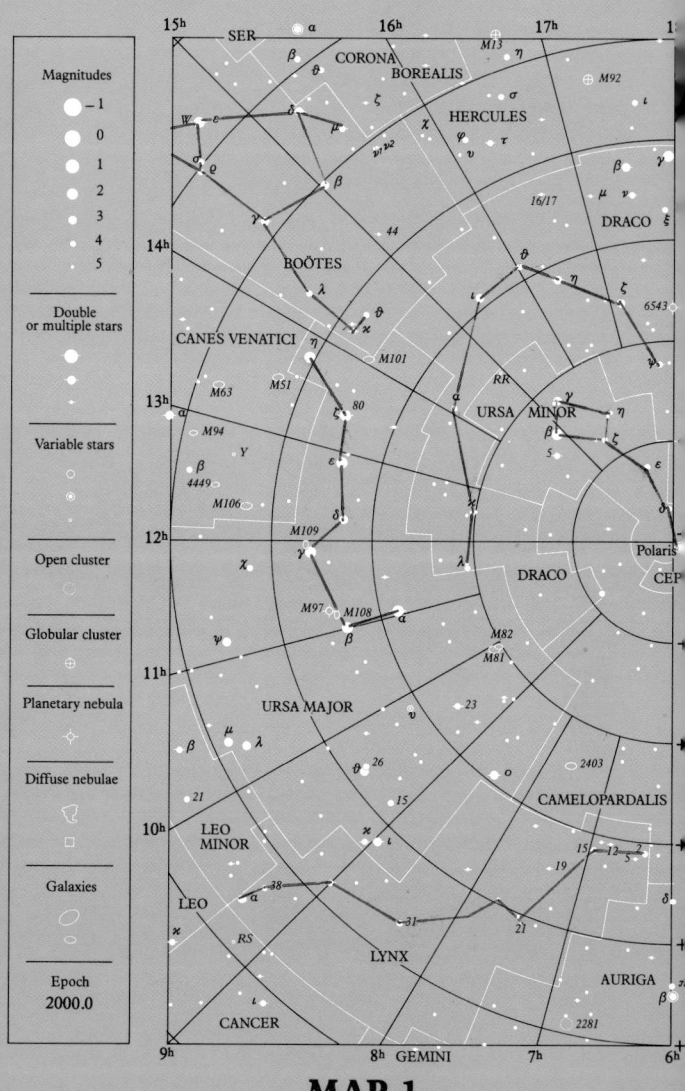

MAP 1

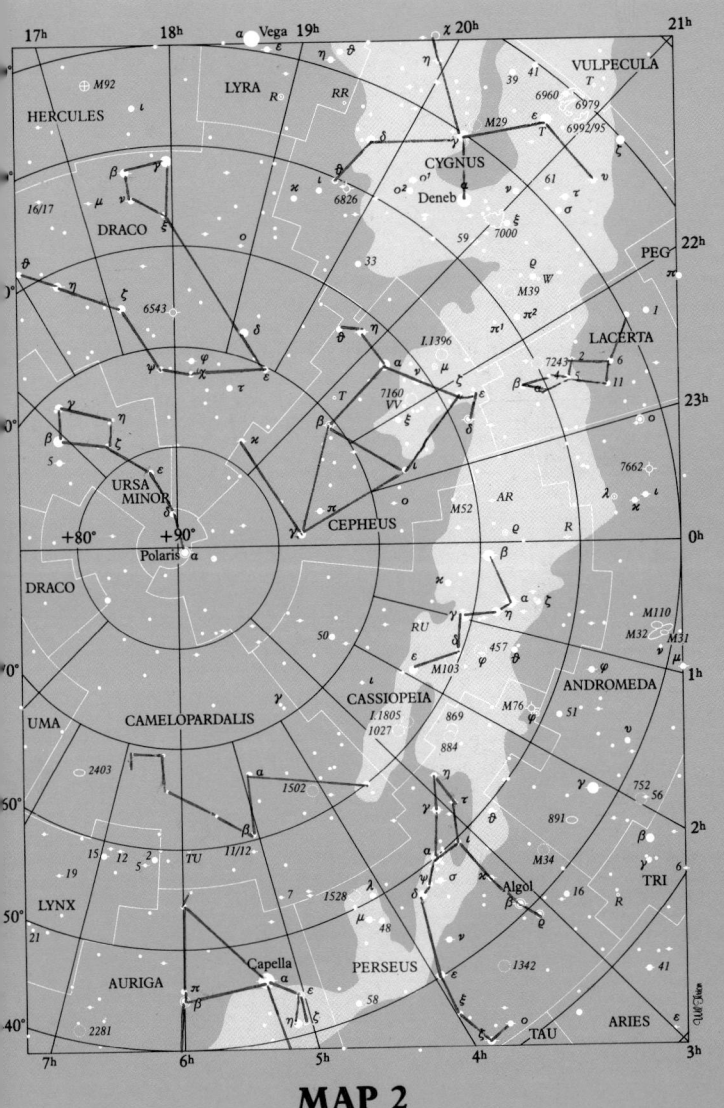

MAP 2

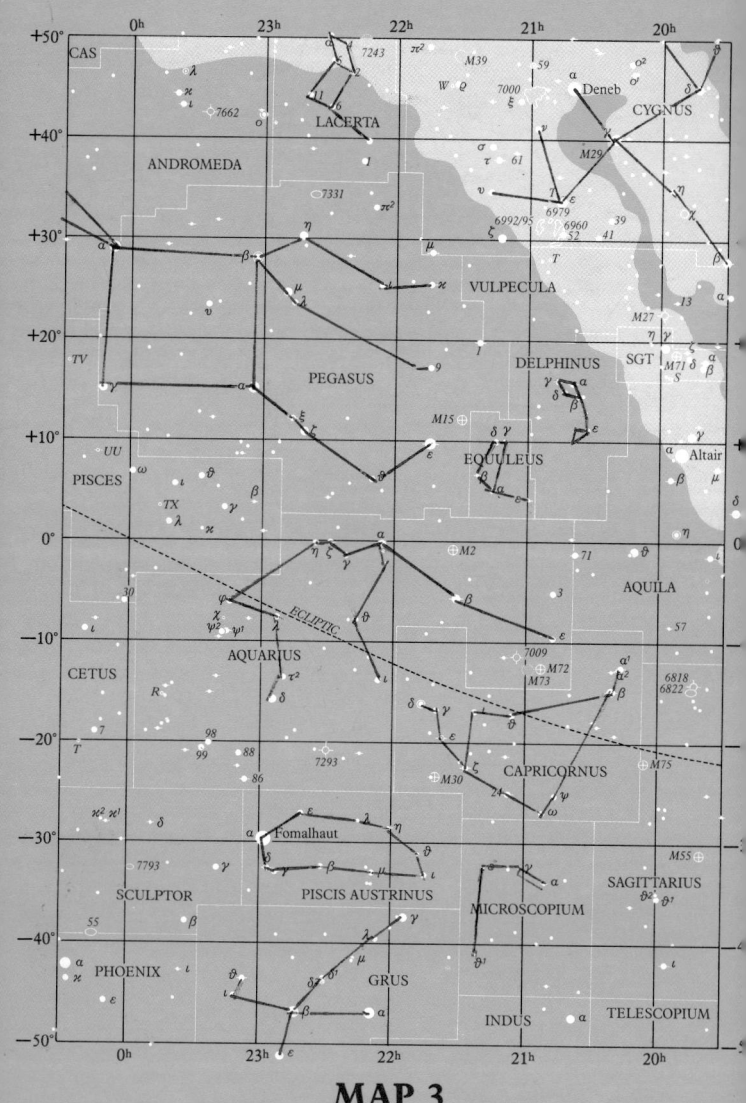

MAP 3

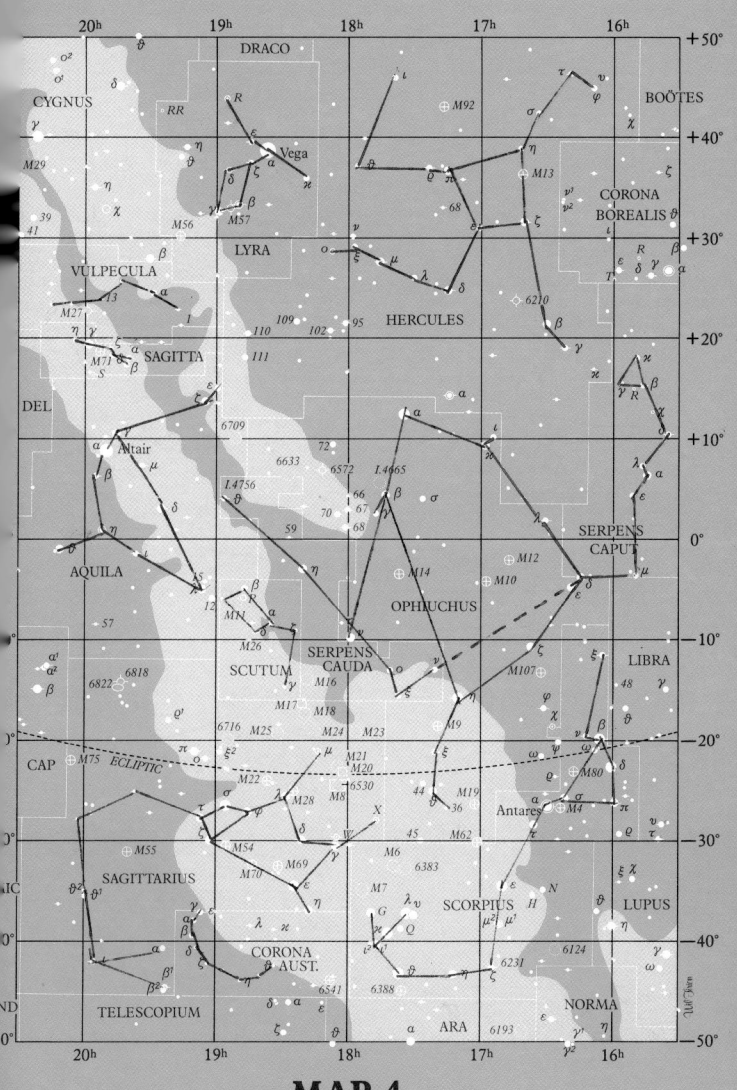

MAP 4

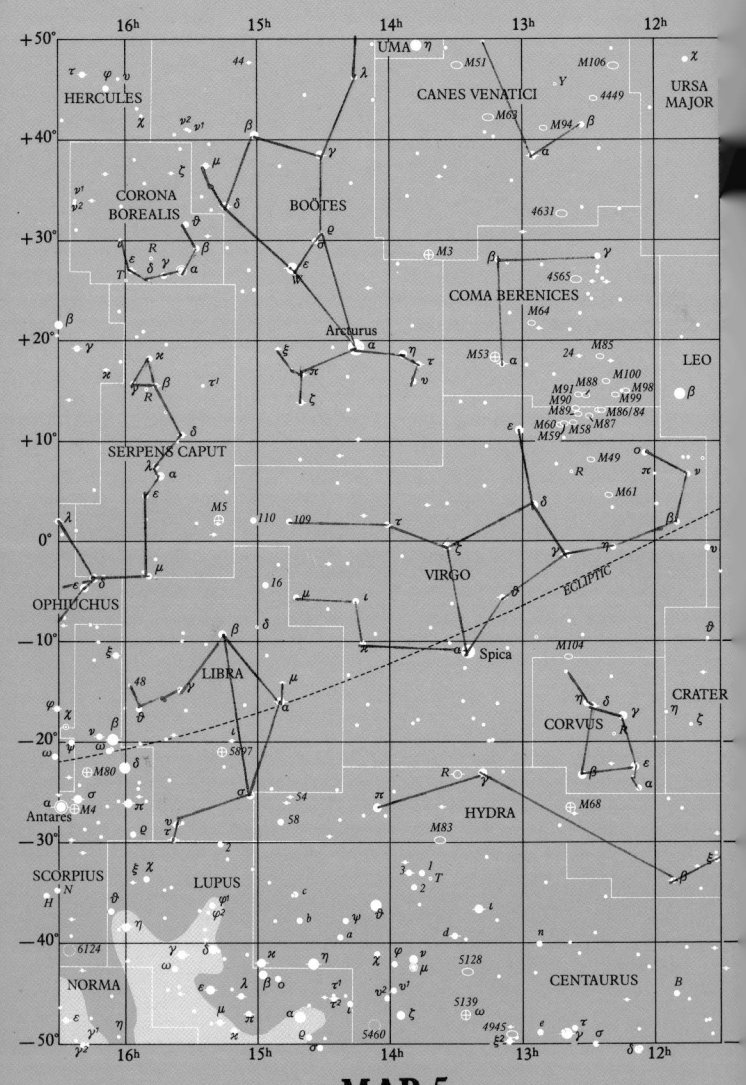

MAP 5

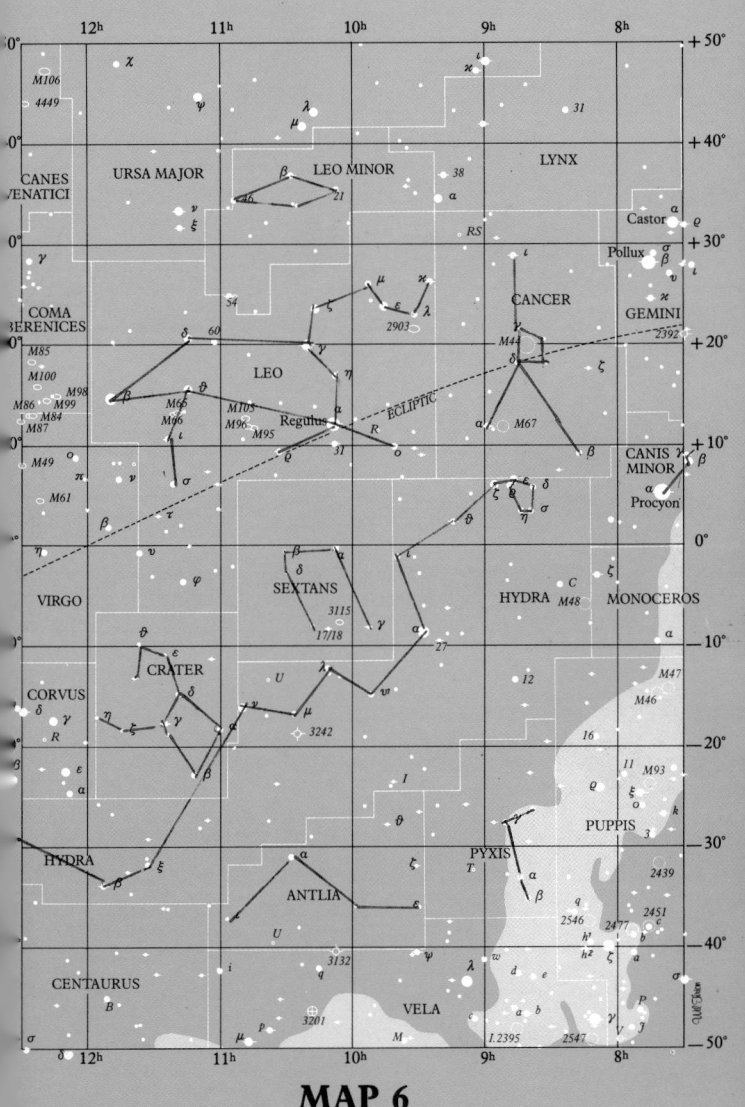

MAP 6

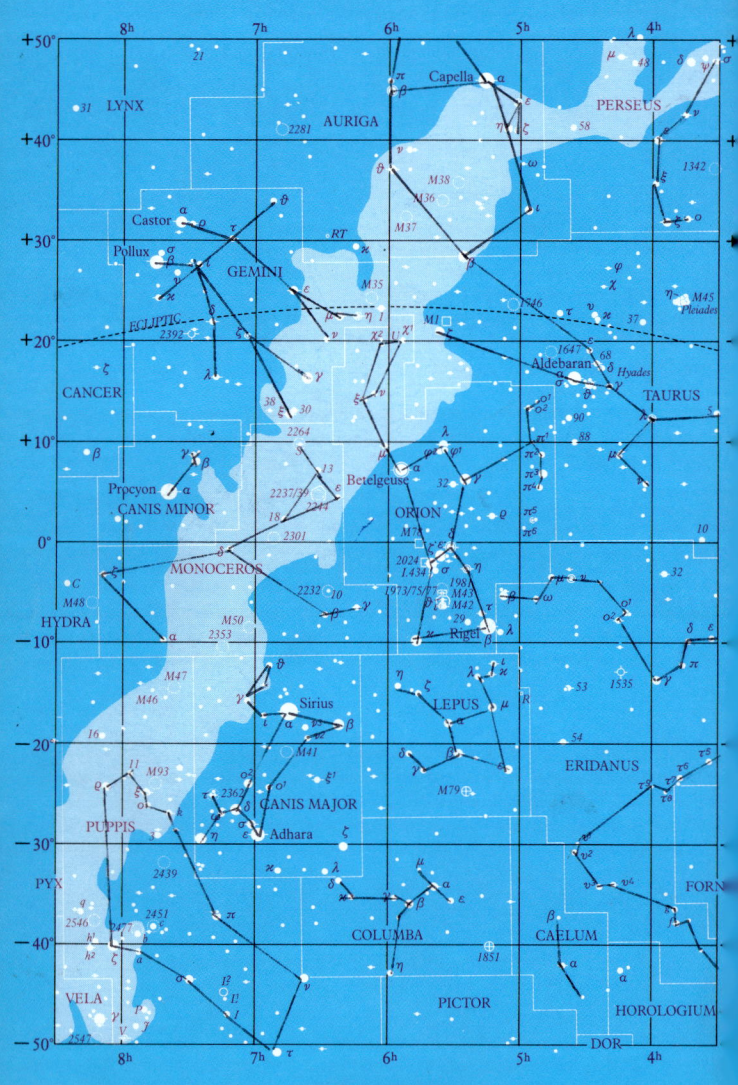

MAP 7

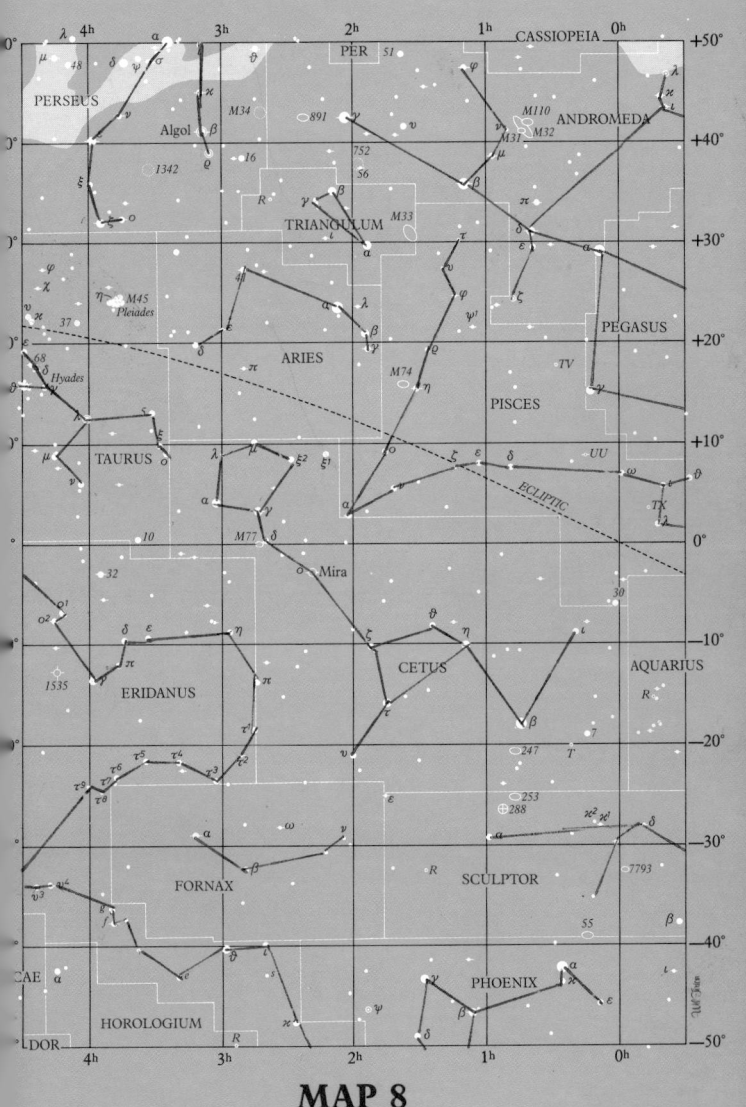

MAP 8

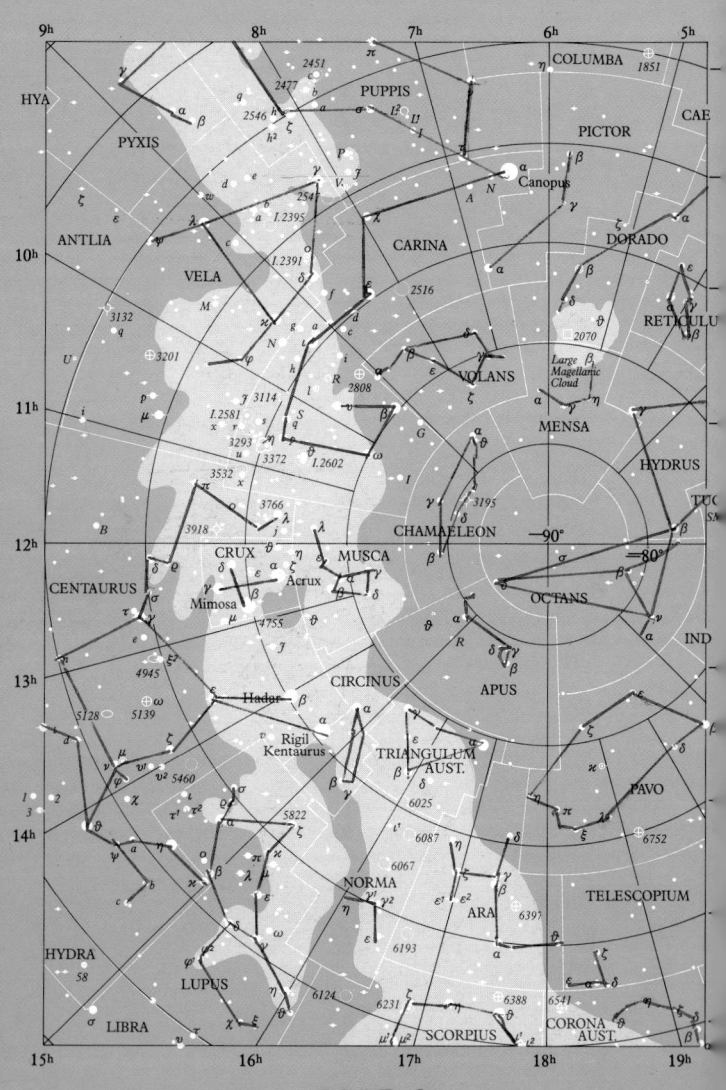

MAP 9

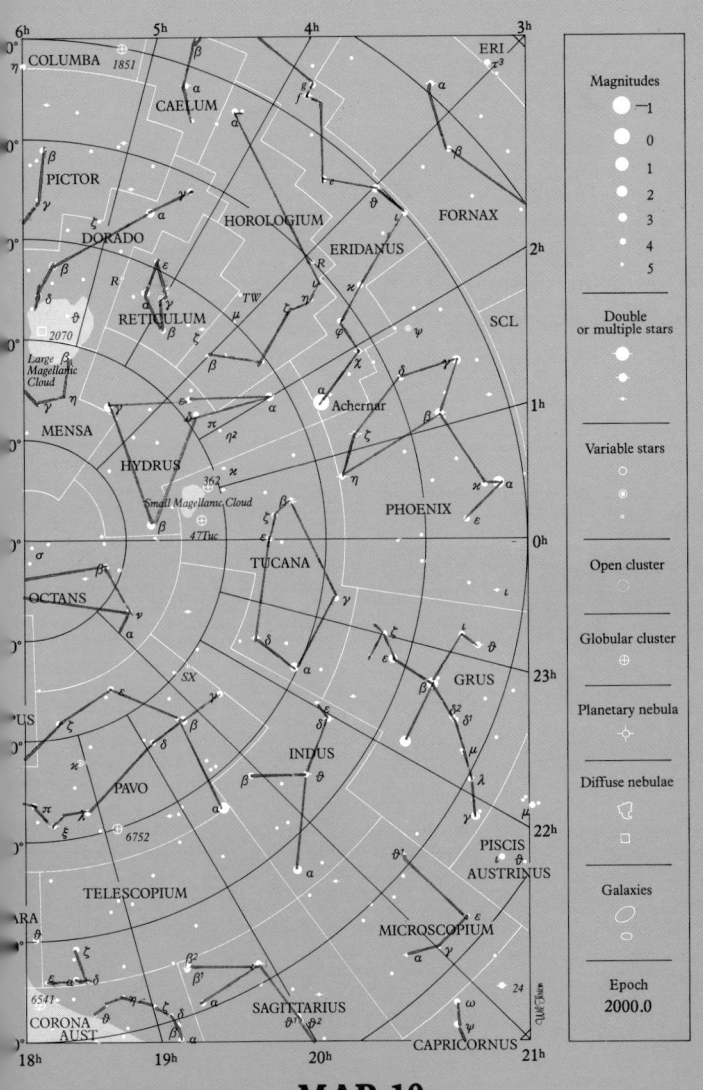

MAP 10

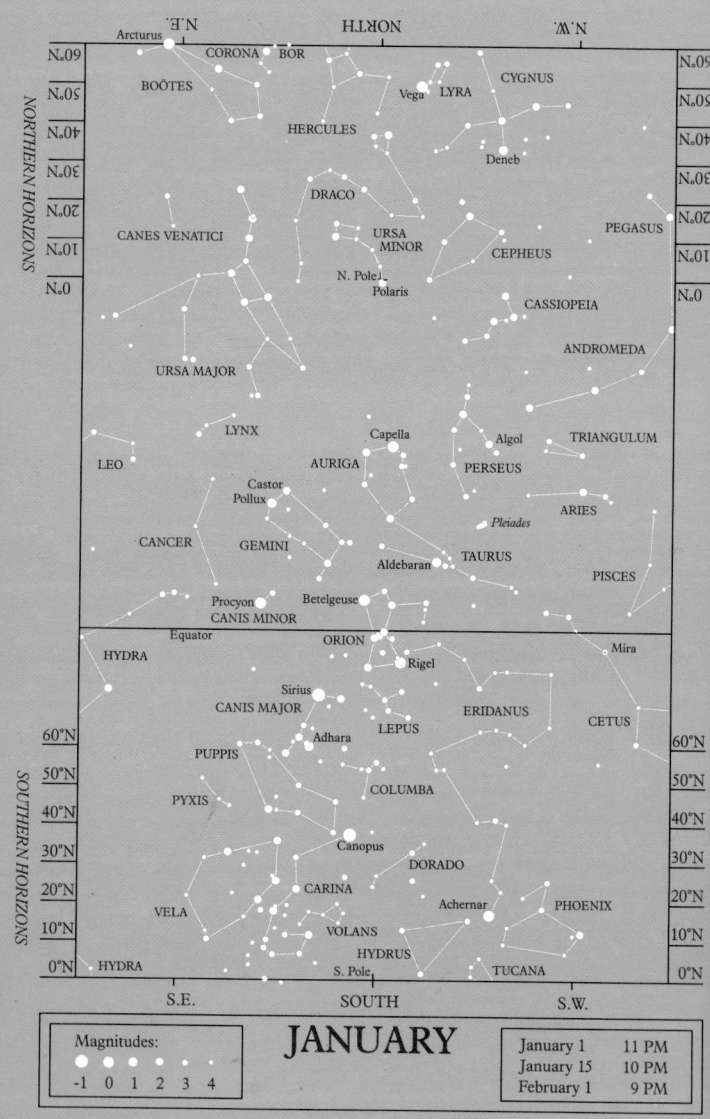

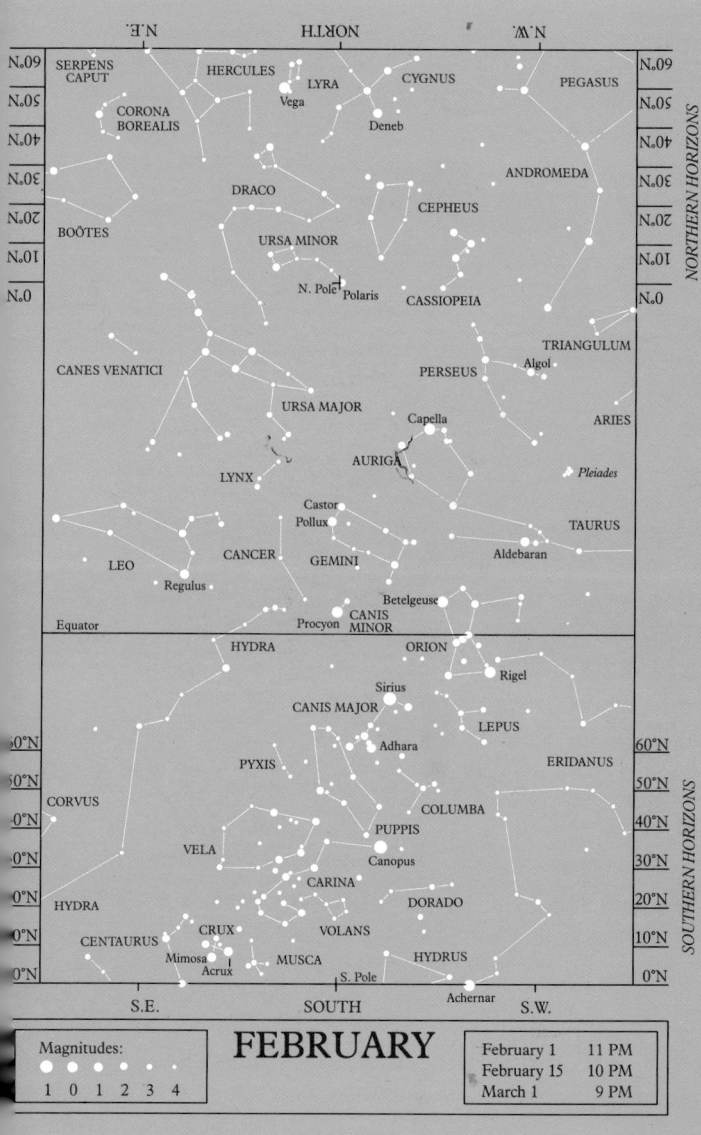

FEBRUARY

Magnitudes:					
1	0	1	2	3	4

February 1 — 11 PM
February 15 — 10 PM
March 1 — 9 PM

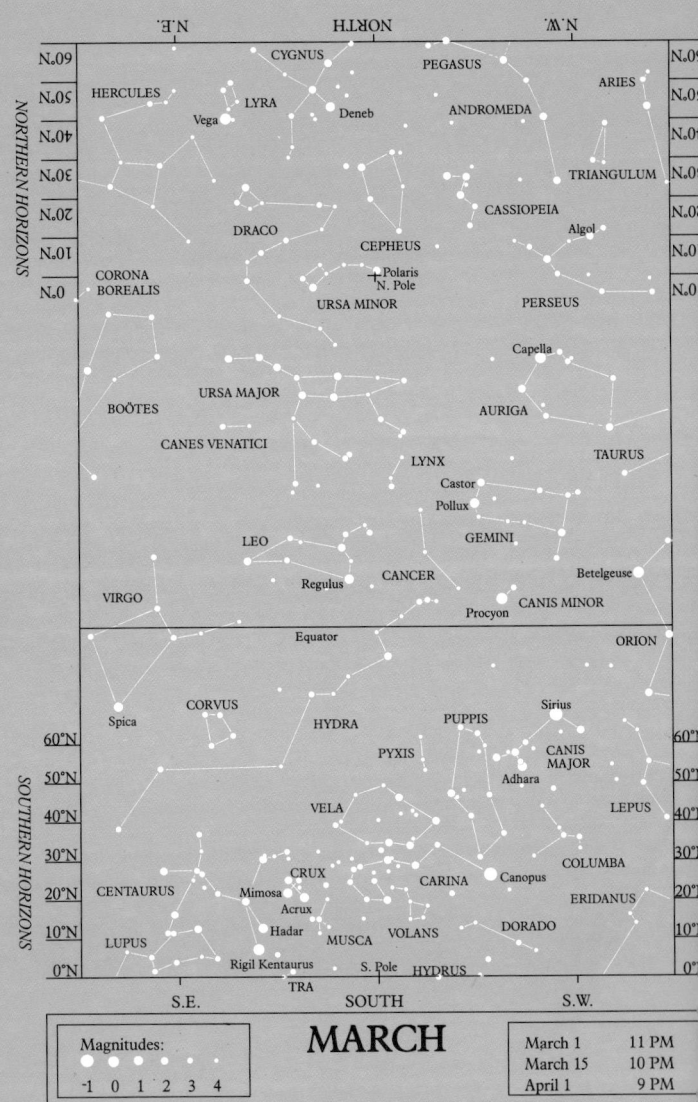

NORTHERN HORIZONS

60°N
50°N
40°N
30°N
20°N
10°N
0°N

60°N
50°N
40°N
30°N
20°N
10°N
0°N

CYGNUS
PEGASUS
ARIES
HERCULES
LYRA
Vega
Deneb
ANDROMEDA
TRIANGULUM
CASSIOPEIA
DRACO
Algol
CEPHEUS
CORONA
BOREALIS
Polaris
N. Pole
URSA MINOR
PERSEUS
Capella
URSA MAJOR
AURIGA
BOÖTES
CANES VENATICI
TAURUS
LYNX
Castor
Pollux
LEO
GEMINI
Regulus
CANCER
Betelgeuse
VIRGO
Procyon
CANIS MINOR

Equator

ORION

SOUTHERN HORIZONS

60°N
50°N
40°N
30°N
20°N
10°N
0°N

60°N
50°N
40°N
30°N
20°N
10°N
0°N

Spica
CORVUS
HYDRA
PUPPIS
Sirius
PYXIS
CANIS
MAJOR
Adhara
VELA
LEPUS
CENTAURUS
CRUX
CARINA
Canopus
COLUMBA
Mimosa
ERIDANUS
Acrux
MUSCA
VOLANS
DORADO
LUPUS
Hadar
Rigil Kentaurus
S. Pole
HYDRUS
TRA

S.E.   SOUTH   S.W.

MARCH

Magnitudes:		March 1	11 PM
		March 15	10 PM
-1 0 1 2 3 4		April 1	9 PM

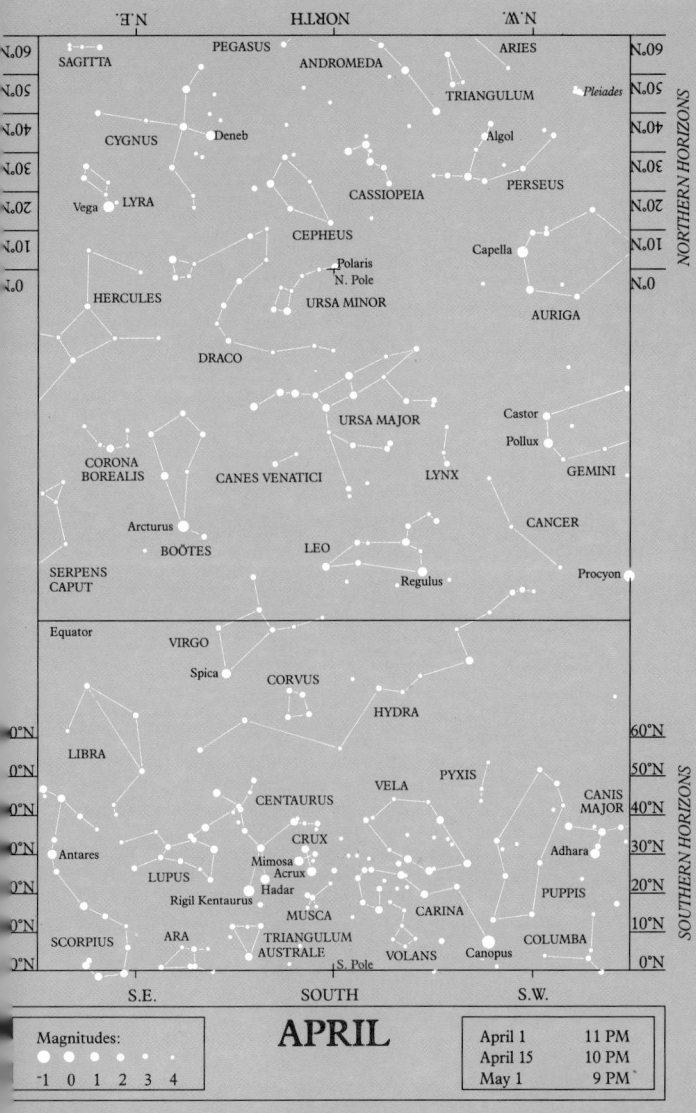

N.E.   NORTH   N.W.

| 60°N | 50°N | 40°N | 30°N | 20°N | 10°N | 0°N |

SAGITTA
PEGASUS
ANDROMEDA
ARIES
TRIANGULUM
Pleiades
CYGNUS
Deneb
Algol
PERSEUS
Vega
LYRA
CASSIOPEIA
CEPHEUS
Capella
HERCULES
Polaris
N. Pole
URSA MINOR
AURIGA
DRACO
URSA MAJOR
Castor
Pollux
GEMINI
CORONA
BOREALIS
CANES VENATICI
LYNX
CANCER
Arcturus
BOÖTES
LEO
Procyon
SERPENS
CAPUT
Regulus

Equator
VIRGO
Spica
CORVUS
HYDRA

| 60°N | 50°N | 40°N | 30°N | 20°N | 10°N | 0°N |

LIBRA
PYXIS
CANIS
MAJOR
CENTAURUS
VELA
Antares
CRUX
Mimosa
Adhara
LUPUS
Acrux
Hadar
PUPPIS
Rigil Kentaurus
MUSCA
CARINA
SCORPIUS
ARA
TRIANGULUM
AUSTRALE
COLUMBA
VOLANS
Canopus
S. Pole

S.E.   SOUTH   S.W.

NORTHERN HORIZONS

SOUTHERN HORIZONS

Magnitudes:	**APRIL**	April 1	11 PM
-1 0 1 2 3 4		April 15	10 PM
		May 1	9 PM

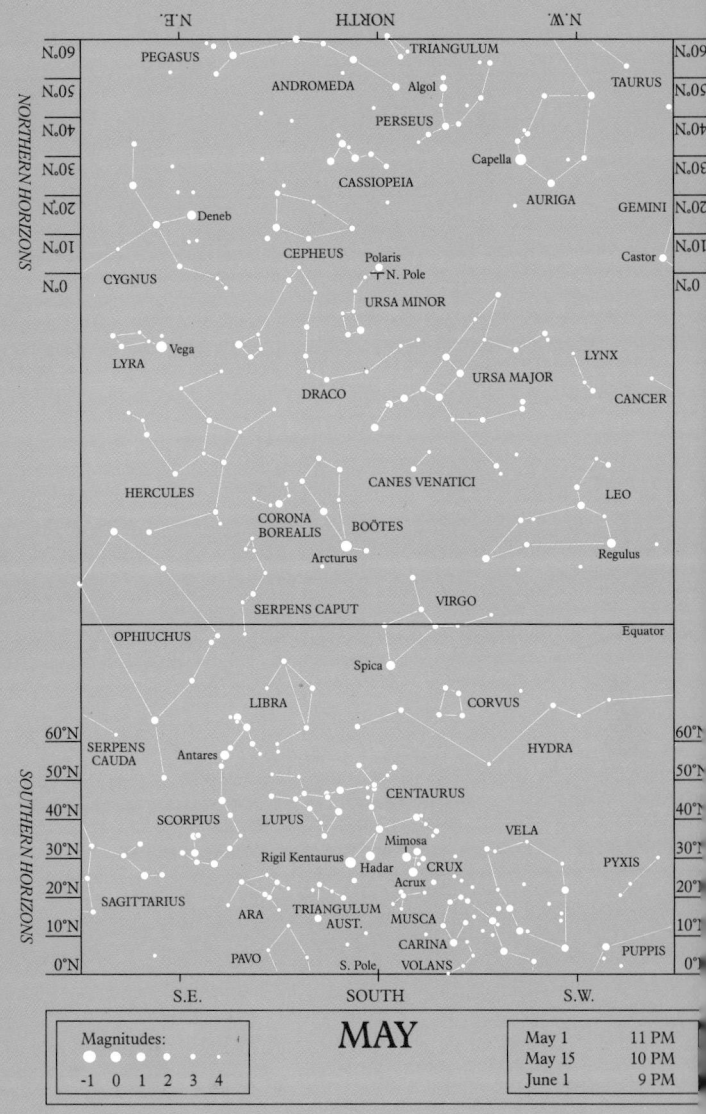

MAY

Magnitudes:						
-1	0	1	2	3	4	

May 1	11 PM
May 15	10 PM
June 1	9 PM

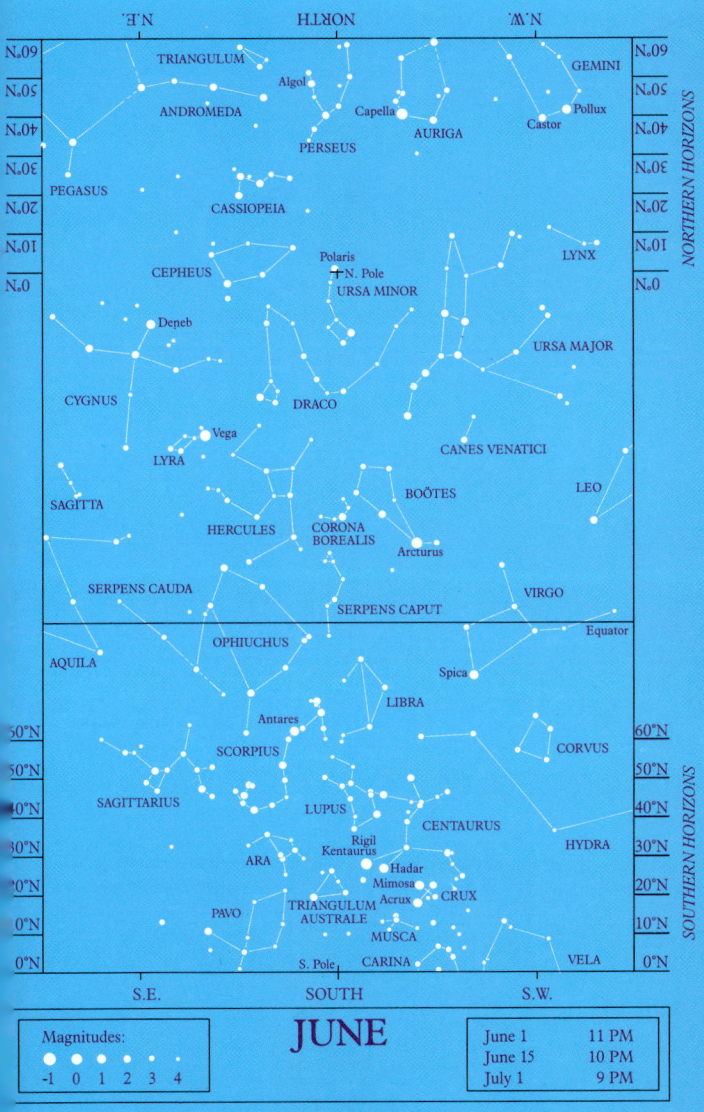

JUNE

Magnitudes:					
-1	0	1	2	3	4

June 1	11 PM
June 15	10 PM
July 1	9 PM

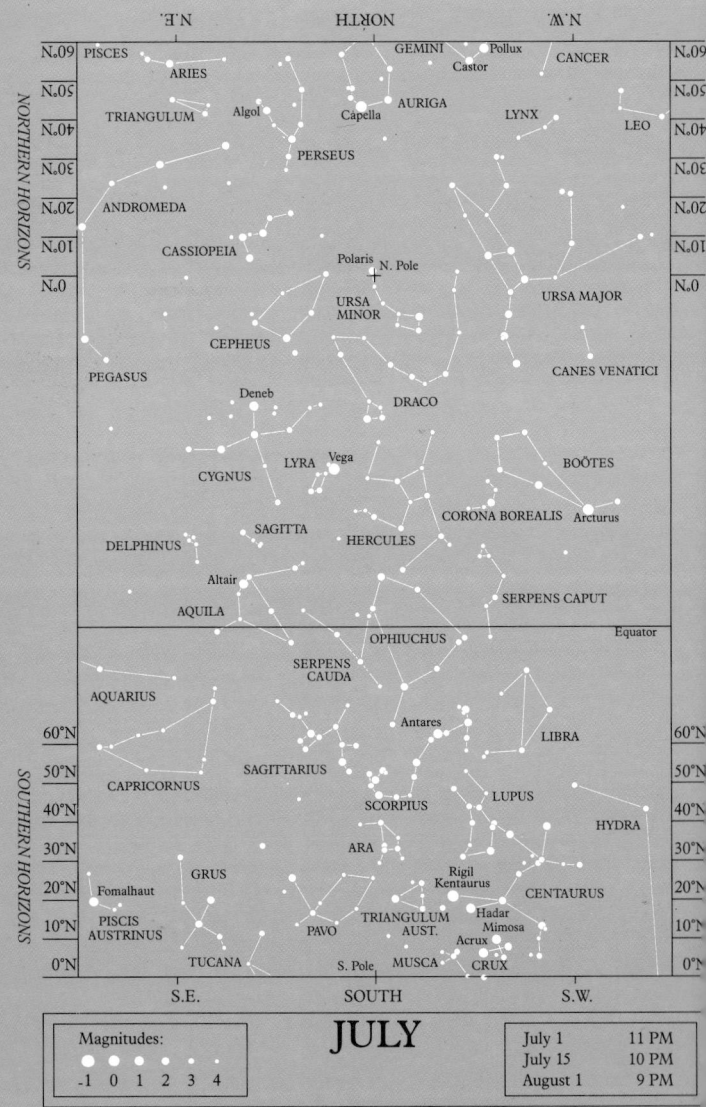

JULY

July 1	11 PM
July 15	10 PM
August 1	9 PM

Magnitudes:
-1 0 1 2 3 4

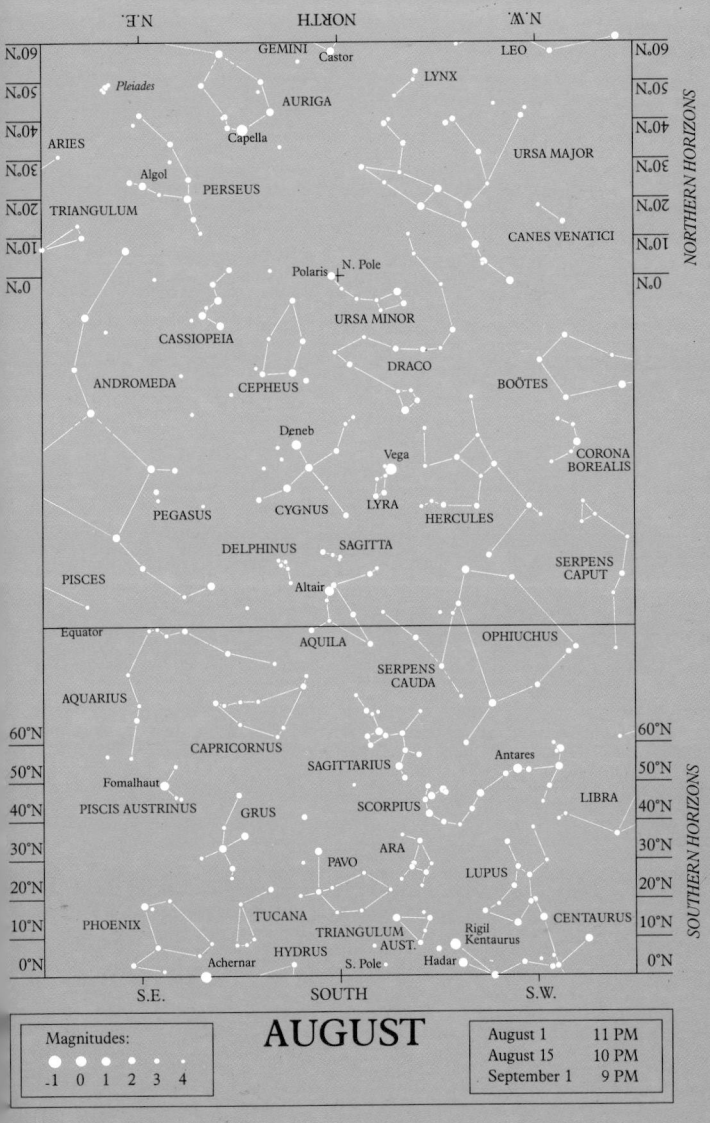

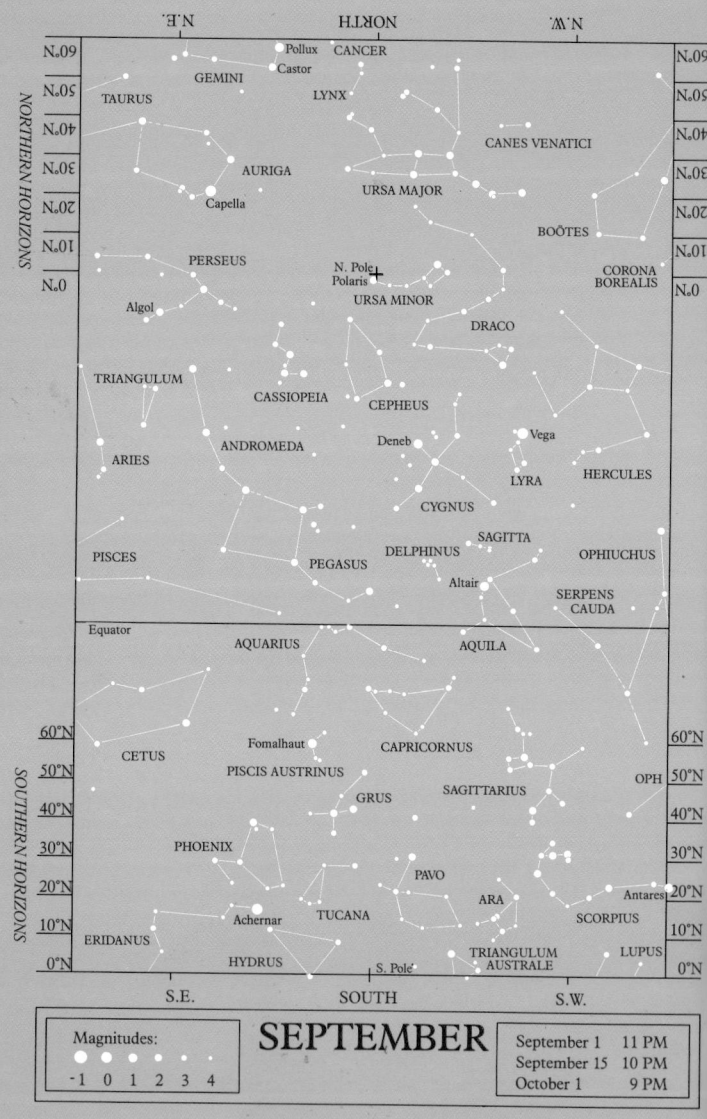

SEPTEMBER

September 1	11 PM
September 15	10 PM
October 1	9 PM

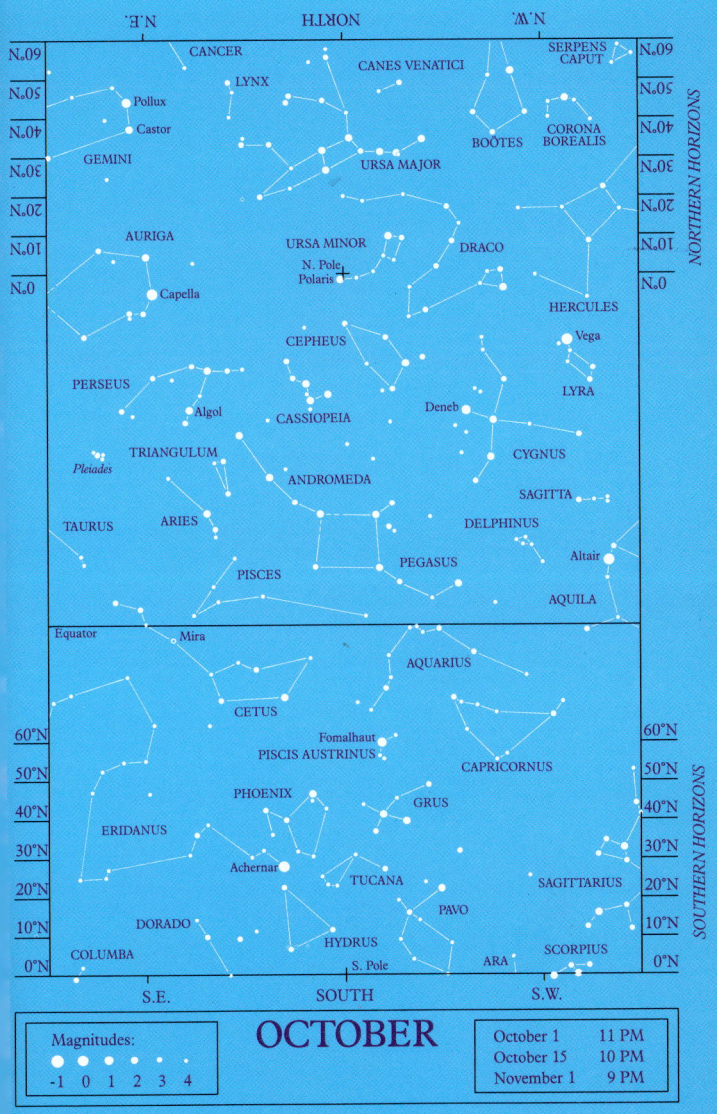

Magnitudes:

-1	0	1	2	3	4

OCTOBER

October 1	11 PM
October 15	10 PM
November 1	9 PM

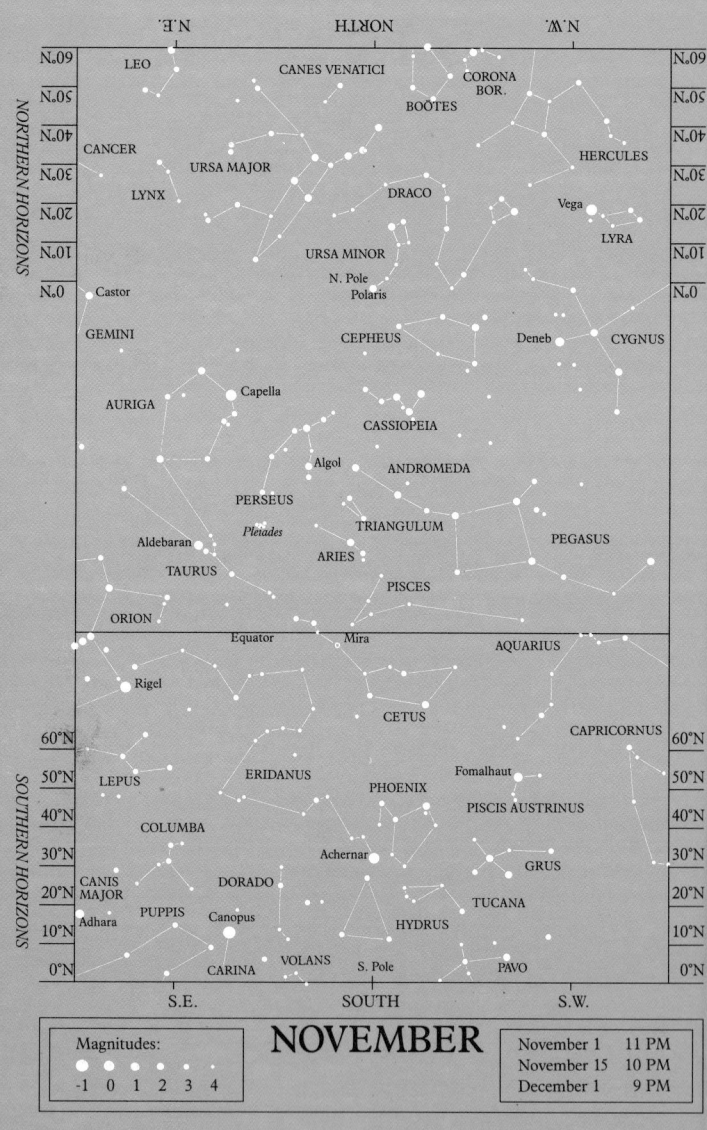

NORTHERN HORIZONS

SOUTHERN HORIZONS

N.E.   NORTH   N.W.

| N.60° | LEO | CANES VENATICI | CORONA BOR. | | N.60° |

BOÖTES

N.50°   N.50°

CANCER   HERCULES

N.40°   N.40°

URSA MAJOR

LYNX   DRACO   Vega

N.30°   N.30°

LYRA

N.20°   N.20°

URSA MINOR

N.10°   N.10°

N. Pole
Polaris

N.0°   Castor   N.0°

GEMINI   CEPHEUS   Deneb   CYGNUS

AURIGA   Capella

CASSIOPEIA

Algol   ANDROMEDA

PERSEUS   TRIANGULUM   PEGASUS

Pleiades

Aldebaran   ARIES

TAURUS   PISCES

ORION

Equator   Mira   AQUARIUS

Rigel   CETUS

CAPRICORNUS

60°N   60°N

Fomalhaut

50°N   LEPUS   ERIDANUS   PHOENIX   PISCIS AUSTRINUS   50°N

40°N   40°N

COLUMBA

30°N   Achernar   GRUS   30°N

DORADO

20°N   CANIS   TUCANA   20°N
MAJOR

10°N   Adhara   PUPPIS   Canopus   HYDRUS   10°N

0°N   CARINA   VOLANS   S. Pole   PAVO   0°N

S.E.   SOUTH   S.W.

Magnitudes:

● ● ● ● · ·
-1 0 1 2 3 4

NOVEMBER

November 1	11 PM
November 15	10 PM
December 1	9 PM

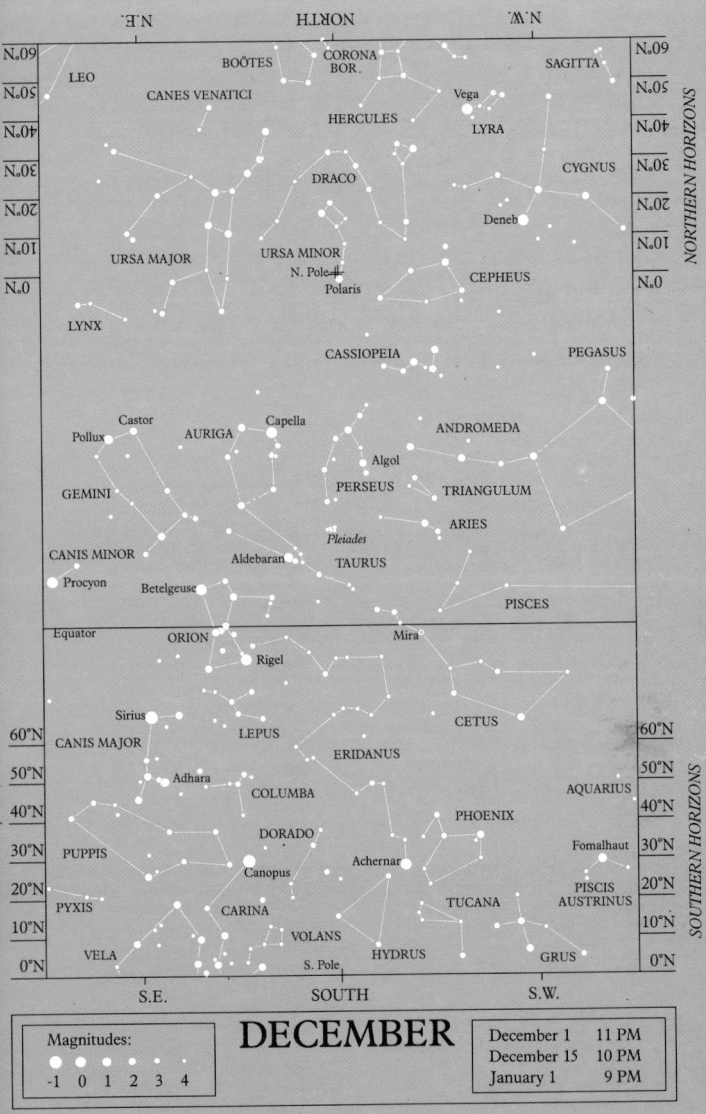

Index

Page numbers in *italic* refer to illustrations or tables.

Prefixes such as the Greek Bayer letters and the Flamsteed numbers of star designations are ignored in the alphabetical ordering; thus γ Andromedae is listed at 'Andromedae', 61 Cygni at 'Cygni', and RS Ophiuchi at 'Ophiuchi'.

There are no references to the notes on interesting objects on pages 144–65.

Acknowledgements

The author wishes to thank the following people for supplying information and for critically reviewing various portions of the manuscript: Geoff Elston, John Fletcher, David Graham, Peter Grego, John Isles, Graham Keitch, Alastair McBeath, Robin Scagell, John Smith and Alan Young.

Picture credits

2, 72/3: David Hardy; 6/7, 20, 22/3, 24, 37, 43, 120, 138, 140/1: © Robin Scagell; 8, 9, 10, 19, 25, 27, 34, 41, 50, 68, 74, 75, 81, 82/3, 94, 96, 102, 103, 106, 107, 114 (top), 115, 130: Julian Baum; 11, 12, 78, 86, 90, 109, 110, 111, 112, 113, 114 (bottom), 166-87: Wil Tirion; 21: © Pierre Neirinck; 38/9, 69, 132/3, 134: © Michael Maunder; 42, 49, 71, 117, 124, 128, 139: © Michael Oates; 44: © Denis Buczynski; 46/7: © Peter Grego; 52-67: John Murray; 76, 85, 87, 89, 92: © David Graham; 77, 135: © Ian Ridpath; 84: © Richard Baum; 98: © George Auckinclose; 100/1: © Roy Easto; 118/9: © Ron Arbour; 121, 122, 123, 124 (inset), 125, 126, 127, 129, 131: © John Lewis; 136, 137: © Paul Sutherland; 142/3: © National Maritime Museum, Greenwich.

Organizations for amateurs

Active amateur astronomers are strongly urged to join an organization. In the UK and Commonwealth the leading amateur organization is the British Astronomical Association (Burlington House, Piccadilly, London W1V 9AG). It has various observing sections and publishes a semitechnical journal every two months. Complete beginners are recommended to join the Junior Astronomical Society (36 Fairway, Keyworth, Nottingham NG12 5DU); despite the name, members of all ages are welcome. The JAS publishes the quarterly magazine *Popular Astronomy* and news circulars, and has observing sections to help newcomers. There are also numerous local societies about which your local library should have information, or you can send an s.a.e for information to the Federation of Astronomical Societies, 8 Merestones Drive, Cheltenham, Gloucestershire GL50 2SS.

Full-page text illustrations

2: *The planet Saturn and its ring system.* 6/7: *Composite photograph of the Pleiades and Hyades star clusters.* 22/3: *An amateur astronomer and her telescope.* 38/9: *The Moon blots out the Sun's disk at the total eclipse of 1976 October 23 over the Indian Ocean, photographed from Zanzibar on ISO 64 film, exposure 0.5 second, 45 mm lens at f/1.7.* 72/3: *The Solar System.* 100/1: *Rich star fields in the constellation Cygnus, photographed from La Palma on ISO 1000 film, exposure 5 minutes, with a 50 mm lens at f/1.9.* 118/9: *The Orion Nebula.* 132/3: *Photographing the solar eclipse of 1979 February 23 from Montana, USA, through binoculars. The people behind the photographer are observing the eclipse through welder's glass.* 142/3: *Constellation figures of the northern hemisphere, drawn by Albrecht Dürer in 1515.*